Post-Translational Modifications in Cellular Functions and Diseases

Edited by Shibo Ying

Published in London, United Kingdom

IntechOpen

Supporting open minds since 2005

Post-Translational Modifications in Cellular Functions and Diseases
http://dx.doi.org/10.5772/intechopen.91081
Edited by Shibo Ying

Contributors
Maria Bermudez, Mariana Velez-Vasquez, Dadnover Vargas-Ibarra, Haoli Ying, Ye Chen, Ruolang Pan, Aysegul Yildiz, Yesim Kaya, Yu Zhang, Shibo Ying, Wu Zhenxing, Xiaofen Mo, Chengbo Lang, Jinjing Luo, Jin Zou

First published in London, United Kingdom, 2021 by IntechOpen
IntechOpen is the global imprint of INTECHOPEN LIMITED, registered in England and Wales, registration number: 11086078, 5 Princes Gate Court, London, SW7 2QJ, United Kingdom
Printed in Croatia

British Library Cataloguing-in-Publication Data
A catalogue record for this book is available from the British Library

Additional hard and PDF copies can be obtained from orders@intechopen.com

Post-Translational Modifications in Cellular Functions and Diseases
Edited by Shibo Ying
p. cm.
Print ISBN 978-1-83969-047-1
Online ISBN 978-1-83969-048-8
eBook (PDF) ISBN 978-1-83969-049-5

Meet the editor

Dr. Shibo Ying is a research professor at Hangzhou Medical College (China). He graduated and obtained his Ph.D. in Applied Life Sciences from Tokyo University of Agriculture and Technology (Japan) in 2011. He was awarded a Japanese government scholarship, and he visited the University of California at Davis (UCD) as an exchange student in 2010. After his graduation, he became a research fellow at the German Cancer Research Center (DKFZ) in Heidelberg (Germany). In 2014, he returned to China and studied at the Zhejiang Academy of Medical Sciences. At present, Dr. Ying acts as a biomedical researcher. He has authored or coauthored over 30 scientific publications. His research interests include molecular mechanisms of post-translational modification and the exploration of their clinical relevance in human diseases.

Contents

Preface

Post-Translational Modification (PTM) is an elaborate process that occurs on a protein after its translation by ribosomes is complete. PTMs of proteins are crucial because they may alter the physical and chemical properties, such as folding, stability, activity, localization, and molecular interaction, and consequently, the functions of the targets. PTMs are mostly catalyzed by special enzymes that recognize specific target sequences in specific proteins. To date, more than 400 different types of PTMs have been currently identified, including phosphorylation, methylation SUMOylation, acetylation, ubiquitination, glycosylation, hydroxylation, ADP-ribosylation, palmitoylation, citrullination, and other novel protein-bound amino acid modifications. PTMs can be mainly grouped into reversible and irreversible categories according to their stability or transience. Most reversible PTMs are usually associated with cellular signal transduction and gene regulation, which are utilized as a switch to control the state of cells, being the resting or proliferating, in normal cells. Some irreversible PTMs are also involved in cell apoptosis, cell cycle, and cell differentiation, which may lead to abnormal pathogenesis or tumorigenesis. There has been an increasing appreciation for the roles of PTMs in a wide variety of cellular functions and diseases.

In this context, Chapters 1 and 2 describe the post-translational regulation of cellular pathways, such as MAPK/ERK and PI3K/AKT signaling pathways in cancers. Chapter 3 focuses on PTMs of histones in mesenchymal stromal cell fate decisions. Chapters 4 and 5 describe the new findings in FEN1 and p300/CBP by PTMs, respectively. In Chapter 6, the role of PRMT1 in tumorigenesis and development is reviewed.

This book is a useful reference for those who are involved in basic and clinical research of PTMs and all other related areas.

Shibo Ying
Hangzhou Medical College,
Hangzhou, China

Chapter 1

Regulation of MAPK ERK1/2 Signaling by Phosphorylation: Implications in Physiological and Pathological Contexts

Dadnover Vargas-Ibarra, Mariana Velez-Vasquez and Maria Bermudez-Munoz

Abstract

Protein phosphorylation represents a rapid and reversible post-translational regulation that enables a fast control of protein activation that play key roles in cell signaling. For instance, Mitogen Activated Protein Kinase (MAPK) pathways are activated upon sequential phosphorylations, resulting in phosphorylation of cytosol and nuclear targets. We focus here on MAPK ERK1/2 signaling that accounts for diverse cellular responses such as cell cycle progression, proliferation, differentiation, senescence, migration, formation of GAP junctions, cell adhesion, cell motility, survival and apoptosis. We review the role of protein phosphorylation in MAPK ERK1/2 activation, in its regulation in time and space and how its dysregulation can lead to tumorigenesis.

Keywords: phosphorylation, cell signaling, MAPK, ERK1/2, kinase, phosphatase, cancer, inhibitors

1. Introduction: cell signaling regulation by phosphorylation

Among post-translational modifications, protein phosphorylation is the most common. Vitellin was the first protein which phosphorylation was discovered, by Phoebus Levene in 1906 [1, 2]. In 1954, Burnett and Kennedy reported the process of enzymatic phosphorylation. Then, Edwin Krebs and Edmond Fischer described how phosphorylation and dephosphorylation can take place and they demonstrated how the process is governed by enzymes [3, 4]. In 1992, the Nobel Prize in Physiology or Medicine was awarded jointly to Edmond H. Fischer and Edwin G. Krebs for their discoveries concerning reversible protein phosphorylation as a biological regulatory mechanism.

Phosphorylation is a reversible protein modification and results from the addition of a phosphate group (PO_4) to the polar group of amino acids. The most common amino acids that are phosphorylated are serine (Ser), threonine (Thr) and tyrosine (Tyr). Although phosphorylation of histidine and aspartate residues can also occur, they are less stable than others. Phosphorylation of a protein can change binding to other proteins: because each phosphate group has two negative charges, phosphorylation can cause a conformational change in the protein by attracting a cluster of positively charged amino acid side chains. This can change the binding of ligands on the protein surface and therefore its activity. On the other hand, the addition of a phosphate group to a protein can be recognized by other proteins having for instance SH2 and PTB domains, that then can attach to phosphorylated proteins such as the cytoplasmatic tail of receptor tyrosine kinases (RTK). Finally, phosphorylation can mask a binding site that otherwise holds two proteins together and then can disrupt this interaction.

Enzymes that catalyze the addition of a phosphate group to a protein are kinases; the reaction is unidirectional because of the large amount of free energy released when the phosphates bonds are broken in ATP to produce ADP. The human genome includes more than 500 protein kinases, and it is estimated that more than one-third of the 10,000 proteins in a typical mammalian cell are phosphorylated at any given time, many with more than one phosphate. Conversely, phosphatases are enzymes that remove a phosphate group from a protein, having the opposite function of kinases. Dephosphorylation has more rapid kinetics than phosphorylation by kinases. The human genome contains more than 200 phosphatases, classified into different families including protein tyrosine phosphatases (PTP), the metal-dependent protein phosphatase PPM, the phosphoprotein phosphatase (PPP) that are pSer/pThr- specific, the dual specificity phosphatase (DUSP) family and the PTEN family of lipid phosphatases [5].

Protein phosphorylation may occur at a single site that primes location for subsequent phosphorylations or directly at multiples sites. Thus, a single protein kinase or multiple kinases may act on the target protein, creating a synchronized cascade of phosphorylations. These events participate in dynamic intracellular signaling that enable cells to respond to extracellular stimuli and to adapt to internal changes. Mitogen-protein activated kinases (MAPK) are conserved kinases in eukaryotes, integrating cell signaling pathways that regulate processes such as cell proliferation, cell differentiation and cell death, from yeast to humans. There are four independent MAPK pathways: MAPK ERK1/2, ERK-5 (also referred to as BMK-1), c-Jun N-terminal kinase (JNK), and p38 signaling families. MAPK modules contain 3-tier kinases that are sequentially activated by phosphorylation. MAPK proteins are designated from upstream to downstream signaling pathway: MAPK kinase kinase (MAPKKK) phosphorylates MAPK kinase (MAPKK); MAPKK phosphorylates and thus activates MAPK. We will focus on MAPK ERK1/2 signaling to illustrate how a particular post-translational modification such as phosphorylation can regulate a signaling pathway and how its dysregulation can be implicated in pathological processes such as tumorigenesis.

2. MAPK ERK1/2 pathway: a cell signaling of sequential phosphorylations

The Extracellular Signal-Regulated Kinases (ERK) have key roles in processes like cell growth, cell proliferation and cell survival. In humans, there are three isoforms of ERK: ERK-1, ERK-2 and ERK-5. Hereon we will concentrate on classical MAPK ERK1/2 to comprehend how this signaling is regulated by phosphorylation.

In the canonical human MAPK ERK1/2 pathway there are three types of MAPKKK (A-Raf, B-Raf and Raf-1 or C-Raf kinases), two MAPKK (MEK1, MEK2) and two MAPK ERK-1, ERK-2. Interestingly, MAPK ERK1/2 signaling is basically regulated by phosphorylations. On the first level, Raf are serine/threonine-protein kinases that phosphorylate human MEK on Ser-218 and Ser-222, producing their activation. The Raf family of kinases includes three isoforms with high homology and a similar domain organization. On the second level, MEK1/2 are dual specificity protein kinases that phosphorylate a threonine and a tyrosine residue in a Thr-Glu-Tyr sequence located in ERK1/2, rendering them active. While human ERK-1 is phosphorylated on Thr-202 and Tyr-204, ERK-2 is phosphorylated on Thr-185 and Tyr-187 residues for activation. Phosphorylation of ERK1/2 by MEK1/2 leads to the rearrangement of several polar contacts, which results in conformational changes in neighboring structural elements (reviewed on [6]). Finally, ERK1/2 are serine/threonine kinases that phosphorylate a wide variety of substrates in different subcellular compartments including the Golgi apparatus, the mitochondrial membrane, the cytoplasm and the nucleus.

MAPK ERK1/2 phosphorylate substrates in a short Pro-X- Ser/Thr-Pro consensus motif (X representing any residue) and interactions with docking sites are important for specificity. Two motifs have been described, the D- and F-motifs, that can cooperate to enhance the substrate affinity of ERK and to set phosphorylation kinetics [7]. ERK1/2 phosphorylate more than 600 proteins, leading to responses such as cell cycle progression, proliferation, cytokinesis, transcription, differentiation, senescence, cell death, migration, formation of GAP junctions, actin and microtubule networks, neurite extension, cell adhesion and motility, survival and apoptosis [8]. To ensure that these cell responses are adaptive to stimuli in space and time, a fine regulation of MAPK signaling is thus necessary. Remarkably, control of ERK1/2 signaling is in part ensured by phosphorylations and dephosphorylations.

3. Regulation of MAPK ERK1/2 by kinases and phosphatases

The MAPK ERK1/2 has at least 3 tiers of regulation: the control of the phosphorylation and thus of the activity of Raf, MEK1/2 and ERK1/2. Additionally, upstream signals from cell receptors to Raf and downstream targets of ERK1/2 play an active role in regulating the MAPK ERK1/2 pathway. Interestingly, mechanisms of MAPK regulation are based partly on the same mechanisms that activate this pathway: phosphorylation events. In this section we specify how phosphorylation can regulate MAPK ERK1/2 signaling from Raf to ERK1/2 by the activity of kinases in feedback signals, and through dephosphorylation by phosphatases.

When RTK are activated by growth factors, their phosphorylated tyrosines enable the coupling of adaptor proteins such as GRB2. This latter binds with SOS, a guanine exchange factor that promotes the activation of Ras. From this level, MAPK ERK1/2 signaling axis exerts feedback regulations through phosphorylations. Growth factor stimulation (like epidermal growth factor EGF) of the cell induces the phosphorylation of four serine residues in a region encompassing three proline-rich SH3-binding sites in the C-terminal domain of SOS1 [9]. These phosphorylation events are realized by ERK1/2 and constitute a negative feedback regulation that leads to a reduction in Ras activation. Kinetic simulation model using parameters collected in living cells found that possibly more than four phosphorylation sites decisively suppress SOS activity [10]. Indeed, SOS1 is also phosphorylated by the ERK1/2 effector ribosomal S6 kinase 2 (RSK-2) on Ser1134 and Ser1161, leading to the recruitment of 14-3-3 and is thus a negative regulation of ERK1/2 activity [11] (**Figure 1** and **Table 1**).

In platelets and nexus ERK1/2 is also activated downstream of the small GTPase Rap1. RasGRP2 is the predominant guanine exchange factor that specifically activates Rap1. RasGRP2, playing a similar role to SOS for Ras, is phosphorylated by ERK1/2 on Ser394 located in the linker region implicated in its autoinhibition. In this case, RasGRP2 phosphorylation results also in a negative feedback loop that determines the amplitude and duration of active ERK1/2 [12]. Moreover, Rap1 is

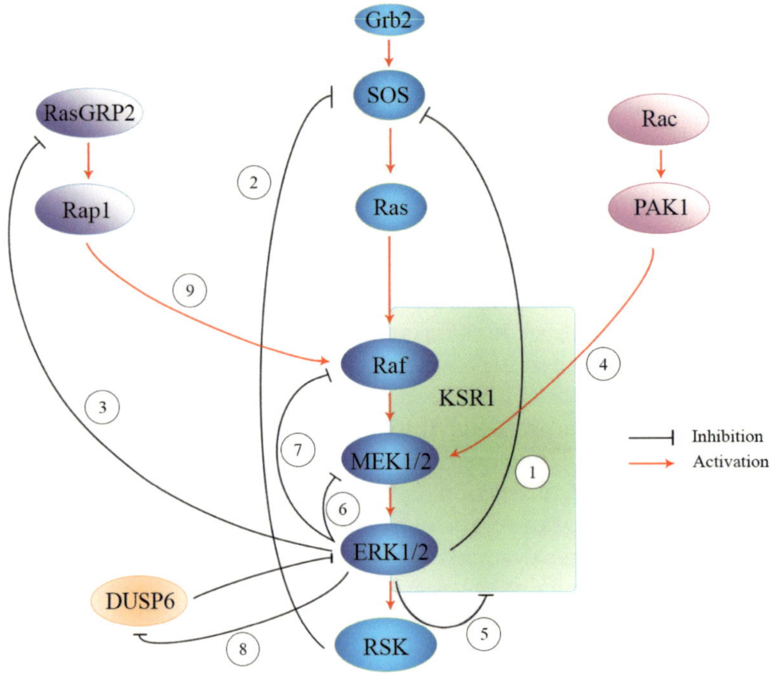

Figure 1.
Representative phosphorylation events leading to activation and feedback signaling in the MAPK ERK1/2 pathway. Phosphorylation constitutes activation (red arrows) or inhibition (black arrows) of proteins of MAPK ERK1/2 signaling. Specific details are provided in ***Table 1****.*

able to phosphorylate and activate B-Raf (but not C-Raf) [23]. Upon cell adhesion and downstream of the small GTPase Rac, the serine/threonine-protein kinase PAK1 phosphorylates the MEK proline-rich sequence (PRS), enhancing its interaction with C-Raf [13] (**Figure 1** and **Table 1**).

Regarding Raf, it has been shown that mitogenic stimuli induce the phosphorylation of C-Raf by ERK1/2 on six residues, needing MEK signaling. Hyperphosphorylation of these sites promotes the subsequent dephosphorylation of C-Raf by PP2A and the return to the inactive state [20]. On the other side, Raf interaction with MEK is also regulated by the inhibitor protein RKIP, which binds to both proteins preventing their physical association. RKIP interferes with the phosphorylation of MEK when bound to C-Raf. Association of RKIP with C-Raf is regulated partly by phosphorylation: phosphorylation of RKIP on serine 153 by PKC or putatively by ERK induces its dissociation from C-Raf [24, 25]. RKIP has then an important role in generating a switch-like behavior of MEK1/2 activity [26].

MEK1/2 is also the target of feedback regulation in the ERK1/2 pathway. Indeed, ERK1/2 phosphorylates MEK1 on Thr292, Thr286 and Thr386, resulting in reduced MEK activity and thus constitutes a negative feedback for MAPK ERK1/2 signaling [18, 19]. Moreover, MEK1 phosphorylation on Thr292 by ERK1/2 interferes with MEK1 binding to ERK2 and reduces MEK1 phosphorylation on S298 by PAK, required for the activation of MEK1 by cell adhesion [13–15] (**Figure 1** and **Table 1**).

Another example of feedback regulation of MAPK ERK1/2 signaling by phosphorylation is the case of the protein scaffold KSR1. In fact, KSR1 can be phosphorylated in Thr256, Thr260, Thr274, Ser320, Ser443, Ser463 by ERK1/2 *in vitro* and depends on MEK1/2 activity. These KSR1 phosphorylations interrupt its association with B-Raf and MEK1/2, drive the release of KSR1 from the plasma membrane, representing then a negative feedback of MAPK ERK1/2 activation [16, 17] (**Figure 1** and **Table 1**).

Number in figure	Protein phosphorylated	Phosphorylation site	Kinase	Type of feedback	Consequence	References
1	SOS1	Ser1132, Ser1167, Ser 1178, Ser 1193	ERK1/2	Negative	Decreased binding affinity of Grb2 to human Sos1	[9]
2	SOS1	Ser1134, Ser1161	RSK	Negative	Facilitates 14-3-3 binding, decreasing MAPK activation	[10]
3	RasGRP2	Ser394	ERK1/2	Negative	Inhibits RasGRP2 ability to activate Rap1, leading to decreased activation of ERK1/2	[12]
4	MEK1/2	Proline-rich sequence (PRS)	PAK1	Positive	Enhances MEK1/2 interaction with C-Raf	[13–15]
5	KSR1	Thr260, Thr274, Ser443	ERK1/2	Negative	Interrupts association of KSR1 with B-Raf and MEK1/2, driving the release of KSR1 from the plasma membrane	[16, 17]
6	MEK1	Thr292	ERK1/2	Negative	Inhibits MEK1 kinase activity towards ERK1/2, interferes with the binding of MEK1 to ERK2 and reduces the ability of PAK to phosphorylate MEK1 on S298 (required for the activation of MEK1 by cell adhesion)	[18, 19]
7	C-Raf	Ser29, Ser43, Ser642, Ser289, Ser296, Ser301	ERK1/2	Negative	Desensitized C-Raf, do not localize to the plasma membrane and do not engage with activated Ras	[20]
8	DUSP6	Ser159, Ser174, Ser197	ERK1/2	Negative	Induces degradation of DUSP6	[21, 22]
9	B-Raf	Ser445	Rap1	Positive	Activation of B-Raf	[23]

Table 1.
Feedback phosphorylation events in MAPK ERK1/2 pathway.

Another regulation of MAPK activity is accomplished by phosphatases that modulate later phases of ERK1/2 signaling. Ser/Thr phosphatases, protein tyrosine phosphatase and dual-specificity Thr/Tyr phosphatases (DUSP) dephosphorylate and thus inactivate ERK1/2. MAP Kinase Phosphatases (MKP) belong to DUSP and represent specific phosphatases that principally regulate MAPK activity in mammalian cells and tissues. While some DUSP dephosphorylate p38, JNK and ERK1/2, others are specific for p38/JNK or for ERK1/2. In this latter case are found cytoplasmic DUSP that inactivate ERK1/2 in the cytoplasm and include DUSP6/MKP-3, a specific phosphatase that binds to ERK1 and ERK2, inactivating them. This specificity is ensured by the fact that the interaction of DUSP6 with ERK1/2 is a requirement for the catalytic activation of the phosphatase through conformational changes [27, 28]. Interestingly, whilst inactivating ERK1/2, DUSP6 is in turn regulated by ERK1/2. Indeed, stimulation with serum or PDGF-B alone can induce a MEK-dependent phosphorylation of DUSP6 on Ser159, Ser174, and Ser197, which is followed by the degradation of the phosphatase by the proteasome [21, 22]. We have shown that another pathway involved in growth factor signaling, the PI3K/mTOR signaling pathway, accounts for a part of the phosphorylation and degradation of DUSP6 induced by serum growth factors. Furthermore, specific agonists of the mTOR pathway, such as amino acids or insulin/IGF-1 are also able to induce the phosphorylation and degradation of DUSP6. Mutagenesis studies identified Ser159 within DUSP6 as the target of the mTOR pathway [29]. Thus, DUSP6 is a point for double MAPK control: the phosphatase exerts a negative regulation for ERK1/2 activity but at the same time, ERK1/2 is able to phosphorylate DUSP6 and then induces its degradation. DUSP6 appears therefore as a spot for fine ERK1/2 signaling regulation in time. Moreover, DUSP6 is a branch-point for the crosstalk between two major signaling pathways induced by growth factors, the MEK/ERK1/2 pathway and the PI3K/mTOR pathway. Notably, both pathways are frequently overactivated in cancer cells. Thus, a regulation of MAPK ERK1/2 signaling in time and space is necessary to warrant cell physiological responses and to avoid aberrant signaling activation that facilitates pathological conditions.

4. Implications of phosphorylation in MAPK ERK1/2 regulation in time and space

MAPK ERK1/2 signaling can determine excluding cell responses such as proliferation and differentiation. Differences in cell responses upon MAPK ERK signaling depend on the regulation of the pathway through protein interactions by scaffolds and through inhibitory and adaptor proteins that enhance, decrease, or redirect the flow of phosphorylation cascades. In this section, we will describe how phosphorylation can be implicated in this type of MAPK ERK1/2 signaling regulation. Scaffold proteins bind to multiple interacting proteins by interconnecting them into a stable complex. This allows the rapid transmission of the signal. Another role of scaffolds is to sequester sets of interacting proteins to limit interactions with other proteins and minimize crosstalk between pathways that some components may share. Scaffold proteins such as KSR1, β-Arrestin, paxillin and IQGAP1 regulate the kinetics, amplitude, and localization of ERK1/2 signaling [30]. Ras-1 suppressor kinase (KSR1) is one of the best characterized scaffold proteins in the ERK1/2 cascade. It has several different domains through which it can interact with C-Raf, MEK1/2, and ERK1/2. In response to growth factors, KSR1 translocates to the plasma membrane where it promotes the activation of MEK1/2 by presenting it to activated Raf. In the absence of stimulus, the ubiquitin-protein isopeptide ligase family member IMP and the 14.3.3 protein prevent the function of KSR1.

Mitogens induce the dephosphorylation of IMP at S392 by protein phosphatase-2A (PP2A) and the degradation of the protein, which is sufficient to allow KSR1 to translocate to the cell membrane [31]. Activated Ras also induces phosphorylation of KSR1 at residues Thr260, Thr274, and Ser443 [16]. Then, while activated Ras prevents the effects of 14.3.3 and IMP that inhibit KSR1 function, it also induces its phosphorylation at Thr274, preparing KSR1 for degradation. KSR1 can then regulate ERK1/2 activation kinetics and influence the biological fate of the cell. The interaction and in particular the synchronization between these processes generates a combinatorial control to modulate both the amplitude and the duration of ERK1/2 activity.

If scaffold proteins play a key role in regulating ERK1/2 signaling in subcellular locations, different factors modulate the strength and the duration of ERK signaling in time: the density of cell surface receptor and its different internalization patterns, the surrounding extracellular matrix and the interaction between kinases and phosphatases. The duration of the signal is critical in determining cell response to ERK1/2 signaling. For instance, long-term ERK1/2 activation can cause differentiation while short-term ERK1/2 activation can lead to cell division. This was initially demonstrated in rat pheochromocytoma PC-12 cells, in which transient activation of ERK1/2 by epidermal growth factor (EGF) or insulin peaks at 5 min and fells back to near-background levels within 15 minutes, and results in cell proliferation. On the other hand, sustained activation of ERK1/2 by nerve growth factor (NGF) persists for more than 60 minutes and induces cell differentiation [32]. This type of cell response according to duration of ERK1/2 signaling has been also reported in fibroblasts, macrophages and T lymphocytes [33–35]. As this type of studies has been made using mainly immunoblotting techniques to monitor ERK1/2 activation dynamics, the use of new approaches gaining spatio-temporal resolution will be of great interest to advance in the understanding of ERK1/2 signaling in time and in subcellular localizations. For example, using Förster Resonance Energy Transfer (FRET)-based ERK biosensors, Keyes et al. showed that EGF induces sustained ERK1/2 activity near the plasma membrane in contrast to the transient activity observed in the cytoplasm and in the nucleus. This supports the concept that the spatial and temporal regulation of ERK1/2 activity is integrated by the cell to control the specificity of signaling [36].

Studies on RTK receptors have shown that their activation kinetics and regulatory mechanisms also play a key role in the activation of the MAPK ERK1/2 pathway. For example, PC-12 cells that express few NGF receptors do not undergo differentiation in response to NGF [37]. Moreover, changing the amount of receptor occupation by decreasing the concentration of agonists alters the duration of ERK1/2 signaling. The rate and degree of receptor internalization also contribute to ERK signaling, not only as a checkpoint for signal termination, but may exhibit additional signaling by the receptor-ligand complex from an internalized cellular location [38].

5. Dysregulation of MAPK ERK1/2 signaling in human cancer

The MAPK/ERK signaling module is considered the most important oncogenic driver of human malignancies [39]. Mutational oncogenic activation of the Ras/Raf/MEK/ERK pathway occurs in a wide variety of cancers concerning approximately 34% of all human cancers. Activation of the ERK1/2 signaling pathway promotes proliferation and has anti-apoptotic effects, increasing tumor invasion and metastasis. The overexpression of the pathway can lead to cell transformation, tumor proliferation, invasion, metastasis, extracellular matrix degradation and tumor angiogenesis. VEGF is an important pro-angiogenic factor and the most powerful pro-vascular endothelial growth cytokine that promotes cell division and vascular construction. The MAPK ERK1/2 signaling pathway can activate transcription factors to enhance the transcription of VEGF, promoting the formation of blood vessels and tumor angiogenesis [40, 41].

Aberrant activation of the Ras/Raf/MEK/ERK pathway may be driven by abnormal receptor kinase activation or by oncogenic mutations of pathway components, leading to tumorigenesis. Overactivation of Ras is observed in approximately 30% of all human cancers but can be higher in some cancers like pancreas cancer (90%), colon cancer (50%) and thyroid cancer (50%) [42, 43]. Mutations in Ras occurs in codons 12, 13, 59 and 61, leading to its constitutive activation. Indeed, mutant oncogenic Ras proteins are insensitive to GTP-catalyzed GTPase hydrolysis activator protein, resulting in a constitutively active GTP-bound Ras. K-Ras and N-Ras are the most common mutated isoforms in human cancer, although H-Ras can also be involved. K-Ras is involved in up to 96% of pancreatic ductal adenocarcinomas, 52% of colorectal carcinomas and 32% of lung adenocarcinomas [44].

Downstream of Ras, Raf can be activated by mutations that mainly affect B-Raf isoform, the most potent activator of MEK1/2 compared with the other Raf isoforms (A-Raf and C-Raf). B-Raf can be mutated in 70% of melanomas, in 36-53% of papillary thyroid cancer, in 30% of ovarian cancer and in 22% of colorectal cancer [45]. The most common mutation of B-Raf is the change of a valine to a glutamic acid in position 600 (V600E). Other B-Raf mutations in cancer are mainly clustered in the activation segment or the so-called glycine-rich loop in B-Raf [46]. Oncogenic mutations of B-Raf lead to hyperactivity of its downstream effectors MEK1/2 and ERK1/2. For cellular transformation to occur, two mutations in Ras/Raf/MEK/ERK1/2 pathway can be needed: for instance, B-Raf and Ras mutations can drive tumorigenesis for colorectal cancer (K-ras G13D; B-Raf G463V), for ovarian cancer (K-ras G13D; B-Raf G463E), and for non-small cell lung cancer (N-Ras Q61K; B-Raf L596V) [45].

Downstream of Raf, MEK1/2 can be highly phosphorylated in colorectal cancer, gliomas, prostate cancer, breast cancer and head and neck cancer [47–51]. Constitutively active mutants of MEK-1 have higher basal activity than the wild-type unphosphorylated MEK. Expression of these mutants in mammalian cells lead to ERK1/2 activation in growth factor-deprived cells, cellular transformation and solid tumor growth in nude mice [33, 52, 53]. If mutant MEK can act as oncogene, its frequency in human cancers appears to be rare [54]. Finally, MAPK ERK1/2 are not frequently mutated. However, some mutations in ERK have been described: ERK2 mutants were identified as rare cancer-associated gain- and loss-of-function gene products: ERK2 D321N, ERK2 E322K, ERK2 L73P, ERK2 S151D and ERK2 D319N [55–60]. While ERK2 D319N has not an increased basal kinase activity, it shows an elevated sensitivity to low levels of signaling *in vivo* [55]. In human cancer cell lines, ERK2 E322K has constitutive phosphorylation [61]. Finally, ERK2 L73P and S151D mutations increase by 8-to-12-fold ERK2 activity alone, and both mutations have a synergetic action that increases by 50-fold ERK2 activity [57]. Moreover, overexpression of ERK2-L73P/S151D can induce growth arrest in prostate cancer cell lines [62]. Although ERK1/2 mutations are rare, mutations that lead to overactivation of RTK, Ras, Raf and MEK can lead to increased ERK1/2 signaling in cancer cells. Downstream of ERK1/2, both cytoplasmic and nuclear targets can be upregulated in tumoral contexts. One of the main cancer-associated ERK substrates is c-Myc, a transcriptional factor that participates in cell cycle progression, becoming an oncogene. Phosphorylation of c-Myc by ERK1/2 due to Ras activation keeps overexpressed this transcriptional factor in various cancers [63]. Other important targets of ERK are the transcriptional factor Elk1, c-Fos and Jun. These two latter were identified as viral oncoproteins and can play a role in tumorigenesis. Mutations that affect MAPK ERK1/2 proteins can then promote protein hyperactivation that induces the cascade of phosphorylation downstream events, favoring cell proliferation, cell transformation and the emergence and progression of tumors. Currently, MAPK inhibitors represent specific target treatments for cancers with overactivation of this cell signaling pathway.

6. MAPK ERK1/2 inhibitors: possibility to regulate cell signaling overactivation

Hyperactivation of Ras/Raf/MEK/ERK signaling pathway in human cancers prompted the development of small molecule inhibitors that target its components for use in cancer therapeutics (**Table 2**). Pharmacological inhibition of Ras has been a major challenge. For instance, the affinity of Ras protein for GTP is extraordinarily high and it is then very difficult to develop a competitive binding strategy. Over the past few years, several groups discovered and developed small molecule Ras modulators using protein structure-guided design approaches [80–82] and exploring SOS as a target for Ras activation [83]. Cysteine-reactive inhibitors that bind to the mutant K-Ras G12C, which is commonly found in cancer, have been developed: SML-8-73-1 and SML-10-70-1 can selectively inhibit K-Ras G12C, changing the nucleotide preference to favor GDP over GTP and thus blocking Ras signaling [69, 84]. These compounds may be used in the future for additional K-Ras mutations.

Sorafenib is an orally available compound that was initially developed as a C-Raf inhibitor and was then identified as a multikinase inhibitor for B-Raf, VEGFR1/2/3, Kit, PDGFR, RET, and Flt3. Sorafenib is currently approved by the FDA for renal and hepatocellular carcinoma for its anti-angiogenic effects [84, 85]. For other cancers like melanoma, sorafenib produced favorable responses in less than 5% of patients in clinical trials [85, 86]. This low response rate can be due to the fact that its activity against B-Raf V600E mutants and wild-type enzymes is low. Subsequent efforts have focused on targeting B-Raf for the treatment of B-Raf mutant melanoma. Vemurafenib and dabrafenib, two B-Raf V600E inhibitors, have achieved benefits in clinical trials [87, 88]. Currently, vemurafenib is approved by the FDA for metastatic and unresectable melanoma with B-Raf V600K mutation [89] and dabrafenib for metastatic melanoma with B-Raf V600K-mutated [84, 88]. Although B-Raf inhibitors have achieved clinical benefit in the treatment of cancer, all ATP-competitive Raf inhibitors including vemurafenib, dabrabenib, and sorafenib can lead to paradoxical activation of the MAPK pathway in wild-type B-Raf cells [90, 91]. Some reports suggest that insensitivity to Raf inhibitors might be due to EGFR-mediated reactivation of MAPK signaling in B-Raf mutant colorectal cancer [92]. Indeed, the combination of EGFR and B-Raf inhibitors block the reactivation of MAPK signaling of B-Raf mutant in colorectal cancer cells and *in vivo* [93]. LGX818, TAK-632 and MLN2480 are other selective B-Raf V600E inhibitors with a very slow inactivation rate, and thus may be beneficial for the treatment of tumors that are resistant to other Raf inhibitors or for the treatment of tumors with Ras mutations [66, 94].

Even though MEK1/2 mutations are rare in human cancers, MEK1/2 have become an attractive drug target because these proteins are downstream of Ras and Raf in the signaling pathway [95]. The first MEK1/2 inhibitor, PD098059, is an allosteric inhibitor that acts on the not-phosphorylated form of MEK1 and mutant MEK1 S217 and S221E [96]. The allosteric MEK inhibitor CI-1040 was

Protein	Mutation	Cancer	Inhibitor	Test/effect/approval	Reference
B-Raf	V600E V600K V226M	Melanoma (66%) Ovarian cancer (35-70%) Thyroid cancer (70%)	Vemurafenib	Approved by the FDA for metastatic and unresectable melanoma with B-Raf V600K mutation	FDA
			Dabrafenib and Trametinib	Approved by the FDA and EMA for melanoma cancer, anaplastic thyroid cancer, NSCLC	FDA, EMA [64]
			LGX818	Approved by the FDA for the treatment of patients with unresectable or metastatic melanoma with B-Raf mutations	FDA [65]
			TAK-632	TAK-632 demonstrates potent antiproliferative effects both on NRAS-mutated melanoma cells and B-Raf-mutated melanoma cells; the combination of TAK-632 and the MAPK kinase (MEK) inhibitor TAK-733 exhibits synergistic antiproliferative effects on these cells	[66]
			MLN2480	In vitro analysis of MLN2480 and TAK-733 (allosteric MEK kinase inhibitor) demonstrates synergistic activity in cell proliferation. In vivo, MLN2480 shows antitumor activity in melanoma, colon, lung, and pancreatic cancer xenograft models	[67, 68]
			Sorafenib	Approved by the FDA for renal and hepatocellular carcinoma	FDA
N-Ras	Q61R Q61L G12D	Melanoma (15- 20%) Myeloid leukemia (30%) Lung cancer (35%)	Ribociclib and Binimetinib	Phase Ib/II trials in patients with locally advanced or metastatic N-Ras mutant melanoma	Clinical trial NCT01781572
	G12V	Thyroid carcinoma (27%)	Trametinib and Palbociclib	Phase I/II trial in patients with solid tumors and with a specific cohort for N-Ras-mutant melanoma	Clinical trial NCT02065063

Protein	Mutation	Cancer	Inhibitor	Test/effect/approval	Reference
K-Ras	G12D G12C G13D	Lung cancer (30%) Colorectal cancer (45%) Pancreatic cancer (90%) Blandder cancer (50%)	SML-8-73-1, SML-10-70-1	SML-10-70-1, a prodrug of SML-8-73-1, inhibits lung cancer A549, H23, and H358 cells	[69]
MEK	S217E S221E	Melanoma (3-8%) Breast cancer (7-9%)	Selumetinib	Approved by the FDA for treatment of pediatric patients aged 2 years and older with neurofibromatosis type 1 (NF1) who have symptomatic, inoperable plexiform neurofibromas; approved by the EMA for the treatment of neurofibromatosis	FDA, EMA
			Trametinib	Approved by the FDA and EMA for treatment of patients with unresectable or metastatic melanoma with B-Raf V600E or V600K mutations	FDA, EMA
			Pimasertib	Has demonstrated potent antitumour activity in human lung, colorectal, melanoma cancer cells and xenograft models; phase I/II clinical trial in patients with locally advanced or metastatic solid tumors	[70–72]
			Cobimetinib	Approved by the FDA and EMA for use in combination with vemurafenib for the treatment of metastatic melanoma	FDA, EMA
			G-573, GDC-0623	In vitro GDC-0623 inhibits cellular proliferation of mutant cancer cells A375 (B-Raf V600E), HCT116 (KRAS G13D), COLO 205 (BRAF V600E), HT- 29 (BRAF V600E), and HCT116 (KRAS G13D). In vivo GDC-0623 causes potent tumor growth inhibition in mouse MiaPaCa-2, A375 and HCT116 xenografts	[73]
			TAK-733	In vitro TAK-733 demonstrates broad activity in most melanoma cell lines; in vivo TAK-733 demonstrates broad antitumor activity in mouse xenograft models of human cancer including melanoma, colorectal, NSCLC, pancreatic and breast cancer	[74]
			Binimetinib(Mek 162)	Approved by the FDA for treatment in combination with LGX818 for patients with unresectable or metastatic melanoma with a B-Raf V600E or V600K mutation; approved by the EMA for the treatment of colorectal carcinoma	FDA, EMA

Protein	Mutation	Cancer	Inhibitor	Test/effect/approval	Reference
ERK 1/2	NA	Melanoma (67%)	Ulixertinib (BVD-523)	In vitro combined Ulixertinib (BVD-523) and VS-5584 treatment causes significant induction of cell death in human pancreatic cancer (HPAC) cells, in pancreatic ductal adenocarcinoma cell lines BxPC-3, MIAPaCa-2, and CFPAC-1. Clinical trials in phase I for tumor advanced pancreatic and other solid tumors cancer and phase II for advanced malignancies harboring MEK or atypical B-Raf alterations.	[75] Clinical trial NCT03454035 and NCT04488003
			GDC-0994	In vitro Ravoxertinib (GDC0994) decreases the viability of lung adenocarcinoma cell lines (A549, HCC827, HCC4006). In vivo GDC0994 results in significant single-agent activity in multiple cancer models, including K-Ras-mutant and B-Raf-mutant human xenograft tumors in mice. Clinical trials phase I for locally advanced or metastatic solid tumors, NSCLC, metastatic colorectal cancer, metastatic NSCLC, metastatic cancers and melanoma	[76] Clinical trials NCT01875705 and NCT02457793
			SCH772984	In vitro SCH772984 results in a G1 arrest in SCH772984-sensitive melanoma cells. In vivo antitumor activity is observed in the K-Ras-mutant pancreatic MiaPaCa model	[77]
			AEZS-134	Synergistic effect of triptorelin, ERK inhibitor AEZS-134 and dual PI3K/ERK inhibitor AEZS-136 in MDA-MB-231 triple-negative breast cancer cells	[78]
			(S)-14 K	In vivo (S)-14 k inhibited tumor growth in mouse xenograft models	[79]

NA: not available; FDA: food and drug administration of USA; EMA: European medicines agency; NSCLC: non-small cell lung cancer.

Table 2.
Available inhibitors for MAPK ERK1/2 proteins.

the first small molecule to enter clinical trials. Although it had antitumor effects, the development of this compound was stopped due to poor bioavailability and lack of efficacy in phase II clinical trials [97]. Other highly selective inhibitors of MEK1 and MEK2 include selumetinib and trametinib [98–100]. This latter prevents Raf-dependent MEK phosphorylation and activation. Other MEK inhibitors in development include pimasertib [101], cobimetinib [102], rafametinib [103], G-573, GDC-0623 [73], TAK-733 [104], RO5126766, RO4987655 [105, 106] and MEK162 [107].

Because there are few ERK1/2 mutations in human cancers, this MAPK has been only considered as a target in 35 clinical trials, compared with more than 300 clinical trials for the inhibition of Raf and MEK. Nonetheless, due to drug resistance resulting from Raf and MEK1/2 inhibitors, ERK1/2 have become an interesting target for inhibiting MAPK ERK1/2 signaling in cancer [46]. ERK1/2 inhibitors can reverse overactivation of the MAPK pathway induced by upstream mutations, including Ras mutations [84, 92, 108]. For instance, MAPK inhibition in B-Raf V600E mutant metastatic melanoma provokes drug resistance and recovery of ERK activity [109, 110]. Interestingly, selective removal of ERK1 or ERK2 *in vitro* can induce melanoma cell death and enhances the action of B-Raf inhibitor [111].

One of the challenges in cancer treatment is developing drug resistance. The mechanisms involved in resistance are complicated and include genetic mutations that occur in target proteins like in MAPK signaling, loss of functions in the control of MAPK signaling feedback, and abnormal tumor suppressor gene alterations [112]. Yet, MAPK inhibitors represent good options for targeting cancer cells with MAPK overactivation or MAPK ERK1/2 mutations. In the future, cell-specific deliverance of MAPK inhibitors to tumoral cells should enhance their efficiency and decrease side effects in patients.

7. Conclusions

MAPK are conserved kinases in eukaryotes, containing 3-tier kinases that are sequentially activated by phosphorylation. This post-translational modification plays an essential role in MAPK ERK1/2 signaling. Not only the activation but also the regulation of this pathway is achieved through the actions of kinases and phosphatases, establishing positive and negative signaling feedbacks. Control of MAPK ERK1/2 signaling in time and space is ensured by proteins such as scaffolds that are themselves regulated by phosphorylation events. Changes in duration of ERK1/2 phosphorylation and thus activity, can result in different cell responses, can result in different cell responses. Thus, a tight regulation of MAPK ERK1/2 signaling is needed to guarantee adaptive cell responses. Aberrant activation of Ras/Raf/MEK/ERK pathway can lead to tumorigenesis and MAPK inhibitors, already in clinical use, represent good options for targeting cancer cells with MAPK overactivation or MAPK ERK1/2 mutations.

Acknowledgements

We thank Helden Natalia Vélez González for assistance in figure design.

Conflict of interest

The authors declare no conflict of interest.

Appendices and nomenclature

DUSP	dual specificity phosphatase
EGF	epidermal growth factor
EMA	European medicines agency
ERK	extracellular signal-regulated kinases
FDA	food and drug administration
FRET	Förster resonance energy transfer
GRB2	growth factor receptor-bound protein 2
KSR1	Ras-1 suppressor kinase
MAPK	mitogen-protein activated kinases
MAPKK	MAPK kinase
MAPKKK	MAPK kinase kinase
MKP	map kinase phosphatase
mTOR	mechanistic target of rapamycin
NGF	nerve growth factor
NSCLC	non-small cell lung cancer
PDGF-B	platelet-derived growth factor subunit B
PDGFR	platelet-derived growth factor
PI3K	phosphatidylinositol 4,5-bisphosphate 3-kinase
PPM	metal-dependent protein phosphatase
PPP	phosphoprotein phosphatase
PRS	proline-rich sequence
PTEN	phosphatidylinositol 3,4,5-trisphosphate 3-phosphatase and dual-specificity protein phosphatase
PTP	protein tyrosine phosphatases
RTK	receptor tyrosine kinase
RSK-2	ribosomal S6 kinase 2
SOS1	son of sevenless homolog 1

Author details

Dadnover Vargas-Ibarra[1†], Mariana Velez-Vasquez[1†] and Maria Bermudez-Munoz[1,2*]

1 Institute of Biology, University of Antioquia, Medellin, Colombia

2 CIDERM, Faculty of Medicine, University of Antioquia, Medellin, Colombia

*Address all correspondence to: olga.bermudez@udea.edu.co

† These authors contributed equally to this work.

References

[1] Levene PA, Alsberg CL. The Cleavage Products of Vitellin. J Biol Chem. 1906 Aug 1;2(1):127-133.

[2] Lipmann FA, Levene PA. Serinephosphoric acid obtained on hydrolysis of vitellinic acid. J Biol Chem. 1932 Oct 1;98(1):109-114.

[3] Burnett G, Kennedy EP. The enzymatic phosphorylation of proteins. J Biol Chem. 1954 Dec;211(2):969-980.

[4] Fischer EH, Krebs EG. Conversion of phosphorylase b to phosphorylase a in muscle extracts. J Biol Chem. 1955 Sep;216(1):121-132.

[5] Chen MJ, Dixon JE, Manning G. Genomics and evolution of protein phosphatases. Sci Signal [Internet]. 2017 Apr 11 [cited 2021 Feb 7];10(474). Available from: https://stke.sciencemag.org/content/10/474/eaag1796

[6] Lavoie H, Gagnon J, Therrien M. ERK signalling: a master regulator of cell behaviour, life and fate. Nat Rev Mol Cell Biol. 2020 Oct;21(10):607-632.

[7] Jacobs D, Glossip D, Xing H, Muslin AJ, Kornfeld K. Multiple docking sites on substrate proteins form a modular system that mediates recognition by ERK MAP kinase. Genes Dev. 1999 Jan 15;13(2):163-175.

[8] Ünal EB, Uhlitz F, Blüthgen N. A compendium of ERK targets. FEBS Lett. 2017 Sep;591(17):2607-2615.

[9] Corbalan-Garcia S, Yang SS, Degenhardt KR, Bar-Sagi D. Identification of the mitogen-activated protein kinase phosphorylation sites on human Sos1 that regulate interaction with Grb2. Mol Cell Biol. 1996 Oct;16(10):5674-5682.

[10] Kamioka Y, Yasuda S, Fujita Y, Aoki K, Matsuda M. Multiple Decisive Phosphorylation Sites for the Negative Feedback Regulation of SOS1 via ERK *. J Biol Chem. 2010 Oct 22;285(43):33540-33548.

[11] Saha M, Carriere A, Cheerathodi M, Zhang X, Lavoie G, Rush J, et al. RSK phosphorylates SOS1 creating 14-3-3-docking sites and negatively regulating MAPK activation. Biochem J. 2012 Oct 1;447(1):159-166.

[12] Ren J, Cook AA, Bergmeier W, Sondek J. A negative-feedback loop regulating ERK1/2 activation and mediated by RasGPR2 phosphorylation. Biochem Biophys Res Commun. 2016 May 20;474(1):193-198.

[13] Eblen ST, Slack-Davis JK, Tarcsafalvi A, Parsons JT, Weber MJ, Catling AD. Mitogen-activated protein kinase feedback phosphorylation regulates MEK1 complex formation and activation during cellular adhesion. Mol Cell Biol. 2004 Mar;24(6):2308-2317.

[14] Coles LC, Shaw PE. PAK1 primes MEK1 for phosphorylation by Raf-1 kinase during cross-cascade activation of the ERK pathway. Oncogene. 2002 Mar 28;21(14):2236-2244.

[15] Slack-Davis JK, Eblen ST, Zecevic M, Boerner SA, Tarcsafalvi A, Diaz HB, et al. PAK1 phosphorylation of MEK1 regulates fibronectin-stimulated MAPK activation. J Cell Biol. 2003 Jul 21;162(2):281-291.

[16] Cacace AM, Michaud NR, Therrien M, Mathes K, Copeland T, Rubin GM, et al. Identification of constitutive and ras-inducible phosphorylation sites of KSR: implications for 14-3-3 binding, mitogen-activated protein kinase binding, and KSR overexpression. Mol Cell Biol. 1999 Jan;19(1):229-240.

[17] McKay MM, Ritt DA, Morrison DK. Signaling dynamics of the KSR1 scaffold

complex. Proc Natl Acad Sci. 2009 Jul 7;106(27):11022-11027.

[18] Brunet A, Gilles Pagès, Jacques Pouysségur. Growth factor-stimulated MAP kinase induces rapid retrophosphorylation and inhibition of MAP kinase kinase (MEK1). FEBS Lett. 1994;346(2-3):299-303.

[19] Rossomando AJ, Dent P, Sturgill TW, Marshak DR. Mitogen-activated protein kinase kinase 1 (MKK1) is negatively regulated by threonine phosphorylation. Mol Cell Biol. 1994 Mar;14(3):1594-1602.

[20] Dougherty MK, Müller J, Ritt DA, Zhou M, Zhou XZ, Copeland TD, et al. Regulation of Raf-1 by direct feedback phosphorylation. Mol Cell. 2005 Jan 21;17(2):215-224.

[21] Jurek A, Amagasaki K, Gembarska A, Heldin C-H, Lennartsson J. Negative and positive regulation of MAPK phosphatase 3 controls platelet-derived growth factor-induced Erk activation. J Biol Chem. 2009 Feb 13;284(7):4626-4634.

[22] Marchetti S, Gimond C, Chambard J-C, Touboul T, Roux D, Pouysségur J, et al. Extracellular signal-regulated kinases phosphorylate mitogen-activated protein kinase phosphatase 3/ DUSP6 at serines 159 and 197, two sites critical for its proteasomal degradation. Mol Cell Biol. 2005 Jan;25(2):854-864.

[23] Carey KD, Watson RT, Pessin JE, Stork PJS. The Requirement of Specific Membrane Domains for Raf-1 Phosphorylation and Activation. J Biol Chem. 2003 Jan;278(5):3185-3196.

[24] Corbit KC, Trakul N, Eves EM, Diaz B, Marshall M, Rosner MR. Activation of Raf-1 signaling by protein kinase C through a mechanism involving Raf kinase inhibitory protein. J Biol Chem. 2003 Apr 11;278(15):13061-13068.

[25] Kwang-Hyun C, Sung-Young S, Hyun-Woo K, Wolkenhauer O, McFerran B, Kolch W. Mathematical Modeling of the Influence of RKIP on the ERK Signaling Pathway. In: Priami C, editor. Computational Methods in Systems Biology. Berlin, Heidelberg: Springer; 2003. p. 127-41. (Lecture Notes in Computer Science).

[26] Shin S-Y, Rath O, Choo S-M, Fee F, McFerran B, Kolch W, et al. Positive- and negative-feedback regulations coordinate the dynamic behavior of the Ras-Raf-MEK-ERK signal transduction pathway. J Cell Sci. 2009 Feb 1;122(Pt 3):425-435.

[27] Camps M, Nichols A, Gillieron C, Antonsson B, Muda M, Chabert C, et al. Catalytic activation of the phosphatase MKP-3 by ERK2 mitogen-activated protein kinase. Science. 1998 May 22;280(5367):1262-1265.

[28] Muda M, Theodosiou A, Gillieron C, Smith A, Chabert C, Camps M, et al. The mitogen-activated protein kinase phosphatase-3 N-terminal noncatalytic region is responsible for tight substrate binding and enzymatic specificity. J Biol Chem. 1998 Apr 10;273(15):9323-9329.

[29] Bermudez O, Marchetti S, Pagès G, Gimond C. Post-translational regulation of the ERK phosphatase DUSP6/MKP3 by the mTOR pathway. Oncogene. 2008 Jun 12;27(26):3685-3691.

[30] Miningou N, Blackwell K. The road to ERK activation: Do neurons take alternate routes? Cell Signal. 2020 Apr 1;68:109541.

[31] Müller J, Ory S, Copeland T, Piwnica-Worms H, Morrison DK. C-TAK1 regulates Ras signaling by phosphorylating the MAPK scaffold, KSR1. Mol Cell. 2001 Nov;8(5):983-993.

[32] Traverse S, Gomez N, Paterson H, Marshall C, Cohen P.

Sustained activation of the mitogen-activated protein (MAP) kinase cascade may be required for differentiation of PC12 cells. Comparison of the effects of nerve growth factor and epidermal growth factor. Biochem J. 1992 Dec 1;288(2):351-355.

[33] Cowley S, Paterson H, Kemp P, Marshall CJ. Activation of MAP kinase kinase is necessary and sufficient for PC12 differentiation and for transformation of NIH 3T3 cells. Cell. 1994 Jun 17;77(6):841-852.

[34] Sharp LL, Schwarz DA, Bott CM, Marshall CJ, Hedrick SM. The influence of the MAPK pathway on T cell lineage commitment. Immunity. 1997 Nov;7(5):609-618.

[35] Whalen AM, Galasinski SC, Shapiro PS, Nahreini TS, Ahn NG. Megakaryocytic differentiation induced by constitutive activation of mitogen-activated protein kinase kinase. Mol Cell Biol. 1997 Apr;17(4):1947-1958.

[36] Keyes J, Ganesan A, Molinar-Inglis O, Hamidzadeh A, Zhang J, Ling M, et al. Signaling diversity enabled by Rap1-regulated plasma membrane ERK with distinct temporal dynamics. eLife [Internet]. 2020 [cited 2021 Feb 14];9. Available from: https://www.ncbi.nlm.nih.gov/pmc/articles/PMC7289600/

[37] Schlessinger J, Bar-Sagi D. Activation of Ras and other signaling pathways by receptor tyrosine kinases. Cold Spring Harb Symp Quant Biol. 1994;59:173-179.

[38] Murphy LO, MacKeigan JP, Blenis J. A network of immediate early gene products propagates subtle differences in mitogen-activated protein kinase signal amplitude and duration. Mol Cell Biol. 2004 Jan;24(1):144-153.

[39] McCubrey JA, Steelman LS, Chappell WH, Abrams SL, Wong EWT, Chang F, et al. Roles of the Raf/MEK/ERK pathway in cell growth, malignant transformation and drug resistance. Biochim Biophys Acta BBA - Mol Cell Res. 2007 Aug 1;1773(8):1263-1284.

[40] Fish JE, Cantu Gutierrez M, Dang LT, Khyzha N, Chen Z, Veitch S, et al. Dynamic regulation of VEGF-inducible genes by an ERK/ERG/p300 transcriptional network. Dev Camb Engl. 2017 Jul 1;144(13):2428-2444.

[41] Song M, Finley SD. Mechanistic insight into activation of MAPK signaling by pro-angiogenic factors. BMC Syst Biol. 2018 Dec 27;12(1):145.

[42] Hobbs GA, Der CJ, Rossman KL. RAS isoforms and mutations in cancer at a glance. J Cell Sci. 2016 Apr 1;129(7):1287-1292.

[43] Malumbres M, Barbacid M. RAS oncogenes: the first 30 years. Nat Rev Cancer. 2003 Jun;3(6):459-465.

[44] Drosten M, Barbacid M. Targeting the MAPK Pathway in KRAS-Driven Tumors. Cancer Cell. 2020 Apr 13;37(4):543-550.

[45] Davies H, Bignell GR, Cox C, Stephens P, Edkins S, Clegg S, et al. Mutations of the BRAF gene in human cancer. Nature. 2002 Jun;417(6892):949-954.

[46] Samatar AA, Poulikakos PI. Targeting RAS-ERK signalling in cancer: promises and challenges. Nat Rev Drug Discov. 2014 Dec;13(12):928-942.

[47] Gee JM, Robertson JF, Ellis IO, Nicholson RI. Phosphorylation of ERK1/2 mitogen-activated protein kinase is associated with poor response to anti-hormonal therapy and decreased patient survival in clinical breast cancer. Int J Cancer. 2001 Jul 20;95(4):247-254.

[48] Gioeli D, Mandell JW, Petroni GR, Frierson HF, Weber MJ. Activation

of mitogen-activated protein kinase associated with prostate cancer progression. Cancer Res. 1999 Jan 15;59(2):279-284.

[49] Gomez-Millan J, Pajares B, Perez-Villa L, Carnero A, Alvarez M, De Luque V, et al. Subcellular localisation of pMEK has a different prognosis in locally advanced head and neck cancer treated with concomitant radiochemotherapy. BMC Cancer. 2016 Oct 28;16(1):829.

[50] Lee SH, Lee JW, Soung YH, Kim SY, Nam SW, Park WS, et al. Colorectal tumors frequently express phosphorylated mitogen-activated protein kinase. APMIS. 2004;112(4-5):233-238.

[51] Mandell JW, Hussaini IM, Zecevic M, Weber MJ, VandenBerg SR. In situ visualization of intratumor growth factor signaling: immunohistochemical localization of activated ERK/MAP kinase in glial neoplasms. Am J Pathol. 1998 Nov;153(5):1411-1423.

[52] Brunet A, Pagès G, Pouysségur J. Constitutively active mutants of MAP kinase kinase (MEK1) induce growth factor-relaxation and oncogenicity when expressed in fibroblasts. Oncogene. 1994 Nov;9(11):3379-3387.

[53] Mansour SJ, Matten WT, Hermann AS, Candia JM, Rong S, Fukasawa K, et al. Transformation of mammalian cells by constitutively active MAP kinase kinase. Science. 1994 Aug 12;265(5174):966-970.

[54] Bansal A, Ramirez RD, Minna JD. Mutation analysis of the coding sequences of MEK-1 and MEK-2 genes in human lung cancer cell lines. Oncogene. 1997 Mar 13;14(10):1231-1234.

[55] Bott CM, Thorneycroft SG, Marshall CJ. The sevenmaker gain-of-function mutation in p42 MAP kinase leads to enhanced signalling and reduced sensitivity to dual specificity phosphatase action. FEBS Lett. 1994 Sep 26;352(2):201-205.

[56] Brenan L, Andreev A, Cohen O, Pantel S, Kamburov A, Cacchiarelli D, et al. Phenotypic Characterization of a Comprehensive Set of MAPK1/ERK2 Missense Mutants. Cell Rep. 2016 Oct 18;17(4):1171-1183.

[57] Emrick MA, Hoofnagle AN, Miller AS, Ten Eyck LF, Ahn NG. Constitutive activation of extracellular signal-regulated kinase 2 by synergistic point mutations. J Biol Chem. 2001 Dec 7;276(49):46469-46479.

[58] Goetz EM, Ghandi M, Treacy DJ, Wagle N, Garraway LA. ERK mutations confer resistance to mitogen-activated protein kinase pathway inhibitors. Cancer Res. 2014 Dec 1;74(23):7079-7089.

[59] Lawrence MS, Stojanov P, Mermel CH, Robinson JT, Garraway LA, Golub TR, et al. Discovery and saturation analysis of cancer genes across 21 tumour types. Nature. 2014 Jan 23;505(7484):495-501.

[60] Ojesina AI, Lichtenstein L, Freeman SS, Pedamallu CS, Imaz-Rosshandler I, Pugh TJ, et al. Landscape of genomic alterations in cervical carcinomas. Nature. 2014 Feb 20;506(7488):371-375.

[61] Arvind R, Shimamoto H, Momose F, Amagasa T, Omura K, Tsuchida N. A mutation in the common docking domain of ERK2 in a human cancer cell line, which was associated with its constitutive phosphorylation. Int J Oncol. 2005 Dec;27(6):1499-1504.

[62] Wu P-K, Hong S-K, Yoon S-H, Park J-I. Active ERK2 is sufficient to mediate growth arrest and differentiation signaling. FEBS J. 2015;282(6):1017-1030.

[63] Sears R, Nuckolls F, Haura E, Taya Y, Tamai K, Nevins JR. Multiple Ras-dependent phosphorylation pathways regulate Myc protein stability. Genes Dev. 2000 Oct 1;14(19):2501-2514.

[64] Odogwu L, Mathieu L, Blumenthal G, Larkins E, Goldberg KB, Griffin N, et al. FDA Approval Summary: Dabrafenib and Trametinib for the Treatment of Metastatic Non-Small Cell Lung Cancers Harboring BRAF V600E Mutations. The Oncologist. 2018;23(6):740-745.

[65] Li Z, Jiang K, Zhu X, Lin G, Song F, Zhao Y, et al. Encorafenib (LGX818), a potent BRAF inhibitor, induces senescence accompanied by autophagy in BRAFV600E melanoma cells. Cancer Lett. 2016 Jan 28;370(2):332-344.

[66] Nakamura A, Arita T, Tsuchiya S, Donelan J, Chouitar J, Carideo E, et al. Antitumor activity of the selective pan-RAF inhibitor TAK-632 in BRAF inhibitor-resistant melanoma. Cancer Res. 2013 Dec 1;73(23):7043-7055.

[67] Cunniff EGC, Zhang J, Chouitar J, Mettetal J, Nakamura K, Arita T, et al. Abstract C146: Combination treatment with the investigational RAF kinase inhibitor MLN2480 and the investigational MEK kinase inhibitor TAK-733 inhibits the growth of BRAF mutant and RAS mutant preclinical models of melanoma and CRC. Mol Cancer Ther. 2013 Nov 1;12(11 Supplement):C146–C146.

[68] Rasco DW, Olszanski AJ, Patnaik A, Espino G, Neuwirth R, Faucette S, et al. MLN2480, an investigational oral pan-RAF kinase inhibitor, in patients (pts) with relapsed or refractory solid tumors: Phase I study. J Clin Oncol. 2013 May 20;31(15_suppl):2547-2547.

[69] Lim SM, Westover KD, Ficarro SB, Harrison RA, Choi HG, Pacold ME, et al. Therapeutic Targeting of Oncogenic K-Ras by a Covalent Catalytic Site Inhibitor. Angew Chem. 2014;126(1):203-208.

[70] Martinelli E, Troiani T, D'Aiuto E, Morgillo F, Vitagliano D, Capasso A, et al. Antitumor activity of pimasertib, a selective MEK 1/2 inhibitor, in combination with PI3K/mTOR inhibitors or with multi-targeted kinase inhibitors in pimasertib-resistant human lung and colorectal cancer cells. Int J Cancer. 2013;133(9):2089-2101.

[71] Delord JP, Houédé N, Awada A, Lebbe C, Lesimple T, Schellens JHM, et al. 616 Pimasertib (MSC1936369B/AS703026), a Selective Oral MEK1/2 Inhibitor, Shows Clinical Activity in Melanoma. Eur J Cancer. 2012 Nov 1;48:190.

[72] Infante JR, Gandhi L, Shapiro G, Burris HA, Bendell JC, Baselga J, et al. Phase lb combination trial of a MEK inhibitor, pimasertib (MSC1936369B), and a PI3K/mTOR inhibitor, SAR245409, in patients with locally advanced or metastatic solid tumors. J Clin Oncol. 2012 May 20;30(15_suppl):TPS3118–TPS3118.

[73] Hatzivassiliou G, Haling JR, Chen H, Song K, Price S, Heald R, et al. Mechanism of MEK inhibition determines efficacy in mutant KRAS-versus BRAF-driven cancers. Nature. 2013 Sep 12;501(7466):232-236.

[74] Micel LN, Tentler JJ, Tan A-C, Selby HM, Brunkow KL, Robertson KM, et al. Antitumor Activity of the MEK Inhibitor TAK-733 against Melanoma Cell Lines and Patient-Derived Tumor Explants. Mol Cancer Ther. 2015 Feb 1;14(2):317-325.

[75] Ning C, Liang M, Liu S, Wang G, Edwards H, Xia Y, et al. Targeting ERK enhances the cytotoxic effect of the novel PI3K and mTOR dual inhibitor VS-5584 in preclinical models of pancreatic cancer. Oncotarget. 2017 May 15;8(27):44295-44311.

[76] Robarge K, Schwarz J, Blake J, Burkard M, Chan J, Chen H, et al. Abstract DDT02-03: Discovery of GDC-0994, a potent and selective ERK1/2 inhibitor in early clinical development. Cancer Res. 2014 Oct 1;74(19 Supplement):DDT02-03-DDT02-03.

[77] Morris EJ, Jha S, Restaino CR, Dayananth P, Zhu H, Cooper A, et al. Discovery of a novel ERK inhibitor with activity in models of acquired resistance to BRAF and MEK inhibitors. Cancer Discov. 2013 Jul;3(7):742-750.

[78] Kwok CW, Treeck O, Buchholz S, Seitz S, Ortmann O, Engel JB. Receptors for luteinizing hormone-releasing hormone (GnRH) as therapeutic targets in triple negative breast cancers (TNBC). Target Oncol. 2015 Sep;10(3):365-373.

[79] Ren L, Grina J, Moreno D, Blake JF, Gaudino JJ, Garrey R, et al. Discovery of Highly Potent, Selective, and Efficacious Small Molecule Inhibitors of ERK1/2. J Med Chem. 2015 Feb 26;58(4):1976-1991.

[80] Burns MC, Sun Q, Daniels RN, Camper D, Kennedy JP, Phan J, et al. Approach for targeting Ras with small molecules that activate SOS-mediated nucleotide exchange. Proc Natl Acad Sci U S A. 2014 Mar 4;111(9):3401-3406.

[81] Maurer T, Garrenton LS, Oh A, Pitts K, Anderson DJ, Skelton NJ, et al. Small-molecule ligands bind to a distinct pocket in Ras and inhibit SOS-mediated nucleotide exchange activity. Proc Natl Acad Sci U S A. 2012 Apr 3;109(14):5299-5304.

[82] Sun Q, Burke JP, Phan J, Burns MC, Olejniczak ET, Waterson AG, et al. Discovery of small molecules that bind to K-Ras and inhibit Sos-mediated activation. Angew Chem Int Ed Engl. 2012 Jun 18;51(25):6140-6143.

[83] Lu S, Jang H, Zhang J, Nussinov R. Inhibitors of Ras–SOS Interactions. ChemMedChem. 2016;11(8):814-821.

[84] Liu F, Yang X, Geng M, Huang M. Targeting ERK, an Achilles' Heel of the MAPK pathway, in cancer therapy. Acta Pharm Sin B. 2018 Jul 1;8(4):552-562.

[85] Roskoski R. Targeting ERK1/2 protein-serine/threonine kinases in human cancers. Pharmacol Res. 2019 Apr 1;142:151-168.

[86] Eisen T, Ahmad T, Flaherty KT, Gore M, Kaye S, Marais R, et al. Sorafenib in advanced melanoma: a Phase II randomised discontinuation trial analysis. Br J Cancer. 2006 Sep;95(5):581-586.

[87] Chapman PB, Hauschild A, Robert C, Haanen JB, Ascierto P, Larkin J, et al. Improved Survival with Vemurafenib in Melanoma with BRAF V600E Mutation. N Engl J Med. 2011 Jun 30;364(26):2507-2516.

[88] Flaherty KT, Puzanov I, Kim KB, Ribas A, McArthur GA, Sosman JA, et al. Inhibition of Mutated, Activated BRAF in Metastatic Melanoma. N Engl J Med. 2010 Aug 26;363(9):809-819.

[89] Bollag G, Tsai J, Zhang J, Zhang C, Ibrahim P, Nolop K, et al. Vemurafenib: the first drug approved for BRAF-mutant cancer. Nat Rev Drug Discov. 2012 Nov;11(11):873-886.

[90] Hatzivassiliou G, Song K, Yen I, Brandhuber BJ, Anderson DJ, Alvarado R, et al. RAF inhibitors prime wild-type RAF to activate the MAPK pathway and enhance growth. Nature. 2010 Mar 18;464(7287):431-435.

[91] Poulikakos P, C Z, G B, Km S, N R. RAF inhibitors transactivate RAF dimers and ERK signalling in cells with wild-type BRAF [Internet].

Vol. 464, Nature. Nature; 2010 [cited 2021 Feb 25]. Available from: https://aplicacionesbiblioteca.udea.edu.co:2598/20179705/

[92] Ahronian LG, Sennott EM, Allen EMV, Wagle N, Kwak EL, Faris JE, et al. Clinical Acquired Resistance to RAF Inhibitor Combinations in BRAF-Mutant Colorectal Cancer through MAPK Pathway Alterations. Cancer Discov. 2015 Apr 1;5(4):358-367.

[93] Corcoran RB, Ebi H, Turke AB, Coffee EM, Nishino M, Cogdill AP, et al. EGFR-mediated re-activation of MAPK signaling contributes to insensitivity of BRAF mutant colorectal cancers to RAF inhibition with vemurafenib. Cancer Discov. 2012 Mar;2(3):227-235.

[94] Stuart DD, Li N, Poon DJ, Aardalen K, Kaufman S, Merritt H, et al. Abstract 3790: Preclinical profile of LGX818: A potent and selective RAF kinase inhibitor. Cancer Res. 2012 Apr 15;72(8 Supplement):3790-3790.

[95] Carlino MS, Todd JR, Gowrishankar K, Mijatov B, Pupo GM, Fung C, et al. Differential activity of MEK and ERK inhibitors in BRAF inhibitor resistant melanoma. Mol Oncol. 2014 May 1;8(3):544-554.

[96] Alessi DR, Cuenda A, Cohen P, Dudley DT, Saltiel AR. PD 098059 Is a Specific Inhibitor of the Activation of Mitogen-activated Protein Kinase Kinase in Vitro and in Vivo(*). J Biol Chem. 1995 Nov 17;270(46):27489-27494.

[97] Lorusso PM, Adjei AA, Varterasian M, Gadgeel S, Reid J, Mitchell DY, et al. Phase I and pharmacodynamic study of the oral MEK inhibitor CI-1040 in patients with advanced malignancies. J Clin Oncol Off J Am Soc Clin Oncol. 2005 Aug 10;23(23):5281-5293.

[98] Davies BR, Logie A, McKay JS, Martin P, Steele S, Jenkins R, et al. AZD6244 (ARRY-142886), a potent inhibitor of mitogen-activated protein kinase/extracellular signal-regulated kinase kinase 1/2 kinases: mechanism of action in vivo, pharmacokinetic/pharmacodynamic relationship, and potential for combination in preclinical models. Mol Cancer Ther. 2007 Aug;6(8):2209-2219.

[99] Lugowska I, Koseła-Paterczyk H, Kozak K, Rutkowski P. Trametinib: a MEK inhibitor for management of metastatic melanoma. OncoTargets Ther. 2015 Aug 25;8:2251-2259.

[100] Yeh TC, Marsh V, Bernat BA, Ballard J, Colwell H, Evans RJ, et al. Biological characterization of ARRY-142886 (AZD6244), a potent, highly selective mitogen-activated protein kinase kinase 1/2 inhibitor. Clin Cancer Res Off J Am Assoc Cancer Res. 2007 Mar 1;13(5):1576-1583.

[101] Kim K, Kong S-Y, Fulciniti M, Li X, Song W, Nahar S, et al. Blockade of the MEK/ERK signalling cascade by AS703026, a novel selective MEK1/2 inhibitor, induces pleiotropic anti-myeloma activity in vitro and in vivo. Br J Haematol. 2010 May;149(4):537-549.

[102] Choo EF, Belvin M, Boggs J, Deng Y, Hoeflich KP, Ly J, et al. Preclinical disposition of GDC-0973 and prospective and retrospective analysis of human dose and efficacy predictions. Drug Metab Dispos Biol Fate Chem. 2012 May;40(5):919-927.

[103] Iverson C, Larson G, Lai C, Yeh L-T, Dadson C, Weingarten P, et al. RDEA119/BAY 869766: a potent, selective, allosteric inhibitor of MEK1/2 for the treatment of cancer. Cancer Res. 2009 Sep 1;69(17):6839-6847.

[104] Dong Q, Dougan DR, Gong X, Halkowycz P, Jin B, Kanouni T, et al.

Discovery of TAK-733, a potent and selective MEK allosteric site inhibitor for the treatment of cancer. Bioorg Med Chem Lett. 2011 Mar 1;21(5):1315-1319.

[105] Leijen S, Middleton MR, Tresca P, Kraeber-Bodéré F, Dieras V, Scheulen ME, et al. Phase I dose-escalation study of the safety, pharmacokinetics, and pharmacodynamics of the MEK inhibitor RO4987655 (CH4987655) in patients with advanced solid tumors. Clin Cancer Res Off J Am Assoc Cancer Res. 2012 Sep 1;18(17):4794-4805.

[106] Rinehart J, Adjei AA, Lorusso PM, Waterhouse D, Hecht JR, Natale RB, et al. Multicenter phase II study of the oral MEK inhibitor, CI-1040, in patients with advanced non-small-cell lung, breast, colon, and pancreatic cancer. J Clin Oncol Off J Am Soc Clin Oncol. 2004 Nov 15;22(22):4456-4462.

[107] Ascierto PA, Schadendorf D, Berking C, Agarwala SS, van Herpen CM, Queirolo P, et al. MEK162 for patients with advanced melanoma harbouring NRAS or Val600 BRAF mutations: a non-randomised, open-label phase 2 study. Lancet Oncol. 2013 Mar;14(3):249-256.

[108] Hatzivassiliou G, Liu B, O'Brien C, Spoerke JM, Hoeflich KP, Haverty PM, et al. ERK Inhibition Overcomes Acquired Resistance to MEK Inhibitors. Mol Cancer Ther. 2012 May 1;11(5):1143-1154.

[109] Long GV, Fung C, Menzies AM, Pupo GM, Carlino MS, Hyman J, et al. Increased MAPK reactivation in early resistance to dabrafenib/trametinib combination therapy of BRAF-mutant metastatic melanoma. Nat Commun. 2014 Dec 2;5(1):5694.

[110] Marampon F, Ciccarelli C, Zani BM. Biological Rationale for Targeting MEK/ERK Pathways in Anti-Cancer Therapy and to Potentiate Tumour Responses to Radiation. Int J Mol Sci [Internet]. 2019 May 23 [cited 2021 Feb 22];20(10). Available from: https://www.ncbi.nlm.nih.gov/pmc/articles/PMC6567863/

[111] Qin J, Xin H, Nickoloff BJ. Specifically targeting ERK1 or ERK2 kills melanoma cells. J Transl Med. 2012 Jan 25;10:15.

[112] Little AS, Smith PD, Cook SJ. Mechanisms of acquired resistance to ERK1/2 pathway inhibitors. Oncogene. 2013 Mar;32(10):1207-1215.

Chapter 2

Post-Translational Regulation of the Activity of ERK/MAPK and PI3K/AKT Signaling Pathways in Neuroblastoma Cancer

Aysegul Yildiz and Yesim Kaya

Abstract

Pathogenesis of cancer is a multi-step process containing a number of cellular alterations such as post-translational dysregulation of intracellular signaling proteins. These alterations control several functions in carcinogenesis such as angiogenesis, metastasis, evading growth suppressors, and sustaining proliferative signaling. Data of various studies has demonstrated that Phosphatidylinositol 3-kinase (PI3K/AKT) and Mitogen-activated protein kinase (ERK/MAPK) pathways are both abnormally activated in many cancer types, including neuroblastoma. ERK/MAPK and PI3K/AKT signaling pathways that are regulated by sequential phosphorylation upon extracellular stimulation have many important functions in cell cycle, migration, proliferation and apoptosis. Besides their aberrant phosphorylation/activation, there is a crosstalk between these two pathways resulting in an anti-apoptotic effect. In this chapter, carcinogenetic abnormalities in post-translational regulation of the activity of ERK/MAPK and PI3K/AKT pathways in neuroblastoma and other cancers will be summarized. In addition, several crosstalk nodes between two pathways will be briefly explained. All these concepts are not only crucial for thoroughly understanding the molecular basis of carcinogenesis but also choosing the appropriate molecular targets for effective diagnosis and treatment.

Keywords: Neuroblastoma, carcinogenesis, intracellular signaling, phosphorylation, ERK/MAPK, PI3K/AKT

1. Introduction

Cancer pathogenesis is mainly characterized by the accumulation of genetic, epigenetic, and post-translational alterations particularly in cellular signaling pathways leading to the manifestation of the cancer hallmarks such as enabling replicative immortality, sustaining proliferative signaling, activating invasion and metastasis, and inducing angiogenesis. Most of these alterations in the signaling pathways are observed on those that control cell growth, proliferation and death, cell fate and motility such as ERK/MAPK and PI3K/AKT pathways [1]. Under normal conditions, cell signaling process works as a regulated cascade, and as a result of these regulated signaling, healthy tissue structure is maintained, and cellular functions are properly performed. However, in case of carcinogenesis, a multistep progress including various abnormalities in epigenetic and post-translational modifications of the components of these signaling pathways (e.g., acetylation, methylation, phosphorylation, ubiquitination, sumoylation etc.) occurs that triggers tumorigenic growth [2].

In carcinogenesis, three types of gene groups, oncogenes, tumor-suppressor genes, and stability genes are the primary sources for oncogenic mutations. As a result of these mutations, genes are over-expressed/silenced, or mutated proteins with dysregulated functions are produced. However, examining the carcinogenesis in detail revealed that mutated genes are not the only responsible for cancer development, and hence focusing on intracellular signaling pathways rather than individual genes is more significant. Several mutations may also be observed in different components of these signaling pathways, and most of these mutations are known to be common for different cancer types [3].

On the other hand, ERK/MAPK and PI3K/AKT signaling pathways have cross talking nodes which post-translationally affect their activity and control many of the important cellular functions such as cellular metabolism, cell growth, division, death, differentiation, and movement. However, this crosstalk becomes severely disturbed in many cancers, including neuroblastoma, resulting in rapid disease progression and poor prognosis [4–6].

Neuroblastoma is one of the most common pediatric cancers that arises from immature sympathetic nervous system precursors and localizes in adrenal gland or sympathetic ganglia [7]. Neuroblastoma tumors have a very high degree of heterogeneity, ranging from more favorable to highly aggressive tumors with high lethality. In neuroblastoma, as in many types of cancer, ERK/MAPK and PI3K/AKT pathways in particular are notable in terms of their contribution to oncogenic transformation and severity of the disease [8, 9].

Therefore, in this chapter, function of ERK/MAPK and PI3K/AKT signaling in different cancers, as well as in neuroblastoma will be summarized. Then their aberrant and oncogenic interaction with each other and with other cellular components will be discussed.

2. ERK/MAPK signaling pathway in cancer

ERK/MAPK is a highly conserved signaling pathway in the evolutionary process that provides signal transduction via Receptor Tyrosine Kinases (RTKs). MAPKs regulate important cellular functions (e.g., cell cycle, proliferation, migration etc.) through phosphorylation of specific serine/threonine regions of target proteins [10, 11]. Four MAPK cascades have been identified in mammalian cells: Extracellular signal-regulated kinase (ERK, classical MAPK), C-Jun N-terminal kinase/stress-activated protein kinase (JNK/SAPK), p38 MAP kinase, and ERK5 [12]. Among them, ERK/MAPK is the most important signaling cascade in tumor development. MAPK enzymes in all eukaryotic cells are found at the junctions of mitogenic stimuli received by different receptors. In response to received stimuli, the signal is transferred intracellularly to the small oncogenic G-protein Ras and then to the Raf (MEK kinase) protein. Activated Raf protein activates MEK1/2 (MAPK/ERK kinase or MAP kinase kinase) signal protein which then phosphorylates and activates ERK1/2 ultimately regulating essential cellular events such as gene expression, mitosis, cell viability, apoptosis, motility, differentiation, and cellular metabolism [13].

In addition to phosphorylation/dephosphorylation, positive and negative regulation of ERK/MAPK signaling involves other types of post-translational modifications as well. Among them, post-translational modifications of Ras protein such as farnesylation and methylation are intriguing for providing fully active Ras. Ras protein activation requires a serial post-translational modifications that allow Ras to localize to the plasma membrane [14]. The first post-translational modification is farnesylation of the Ras carboxylterminal peptide CAAX through covalent binding of a farnesyl isoprenoid lipid by cytosolic farnesyltransferase enzyme. Then, Ras converting enzyme 1 (Rce1) cleaves AAX tripeptide to generate a free cysteine residue. Finally, a methyl group is covalently attached to this cysteine residue by isoprenylcysteine-O-carboxyl methyltransferase (ICMT) to facilitate the transfer of Ras to the plasma membrane [15].

On the other hand, ERK/MAPK signaling may be downregulated by post-translational modifications including ubiquitination, SUMOylation, and acetylation/deacetylation reactions, In a study, it was shown that Ras- or MEK-mediated ERK activation is attenuated by SIRT1 through deacetylation of the dual specific phosphatase for MAPK, MKP1 which results in the inhibition of cellular proliferation and transformation. They revealed that binding affinity of deacetylated MKP1 to ERK is increased subsequently leading to the inactivation of ERK [16].

In addition, ERK1/2 protein may be inactivated by its degradation through Ubiquitin-Proteasome System (UPS). This process is mediated by an upstream MAP3K, MEKK1 bearing ubiquitin ligase activity, that triggers ERK ubiquitination [17, 18]. Another member of ERK/MAPK pathway, c-Raf, is also degraded by the UPS under certain conditions. Hsp90 is a chaperone responsible for the stability and function of c-Raf protein, and Hsp90 degradation promotes destabilization and degradation of c-Raf by UPS. In a study, it was identified that the antitumor effect of the antibiotic benzoquinone ansamycin geldanamycin is a result of its binding to Hsp90 which triggers its degradation [19].

Fırthermore, the ERK/MAPK signaling may be downregulated by SUMOylation of MEK (MAPK–ERK kinase). SUMOylated MEK loses its interacting ability with ERK which ends up with the blocking of ERK activation. Oncogenic Ras prevents this process in carcinogenesis by disrupting the SUMO E3-ligase activity of MEKK1 MAPKKK [20].

The ERK/MAPK signaling pathway is one of the main oncogenic pathways and overactivated approximately 30% of human tumors [21]. The ERK/MAPK pathway, particularly when activated by growth factors and mitogens has the strongest correlation with cancer. Carcinogenic abnormalities including over-expressing/activating mutations of RTKs, constant production of activating ligands, and Ras/Raf mutations trigger the continuous activation of the ERK/MAPK pathway which indicates that carcinogenic dysregulation of this pathway may occur at different levels (**Table 1**) [24].

There is a large body of evidence describing the contribution of ERK/MAPK signaling to cancer progression. In a study, it was shown that the expression of MKP-1 which is the negative nuclear regulator of ERK/MAPK signaling is increased in normal ovarian surface epithelium and benign cystadenomas compared to invasive carcinomas and tumors with low malignancy potential and borderline tumors. The level of MKP-1 expression in tumor tissues of patients with stage III/IV disease was found to be significantly lower compared to patients with stage I/II disease which is in contrast to the results indicating a significantly higher expression of phosphorylated-ERK1/2 (p-ERK1/2) in stage III/IV tumors compared with that in stage I/II tumors. These data point out to a negative correlation between MKP-1 and p-ERK1/2 expression in the same ovarian cancer tissue which emphasizes the significance of the abnormal expression of MKP-1 and its effect on ERKs phosphorylation in the development of ovarian cancer [25].

Moreover, in another study with colon tissue, they showed that in colon cancer, tubular adenoma, and villous adenoma, MEK phosphorylation rates were 76, 30

Type of mutations in MAPK signaling pathway	Rate of mutations in different cancers
EGFR **over-expression**	Most carcinomas (>50%)
ERBB2 **over-expression**	Breast (30%)
	Pancreas (90%)
	Lung adenocarcinoma (35%) (non-small cell)
	Thyroid; follicular (55%)
	Thyroid; undifferentiated papillary (60%)
RAS **mutation**	Seminoma (45%)
	Melanoma (15%)
	Bladder (10%)
	Liver (30%)
	Kidney (10%)
	Myelodysplastic syndrome (40%)
	Acute myelogenous leukemia (30%)
BRAF **mutation**	Melanoma (66%)
	Colorectal (12%)
MEK **or** *ERK* **mutation**	Melanoma (3–8%)
	Colorectal (3%)

Table 1.
Mutation rates in MAPK signaling pathway of different cancers [22, 23].

and 40%, respectively. However, the phosphorylation of MEK in normal colonic mucosal cells was scarcely detectable [26]. In addition, in a study examining ERK/MAPK pathway's function in cellular growth and differentiation in colon carcinoma of mice by Sebolt-Leopold et al., oral intake of MEK inhibitor provide inhibition of tumor growth in the rate of 80% [27].

As in other types of cancer, aberrant signaling of ERK/MAPK pathway is crucial for neuroblastoma cancer since it leads to reduced therapeutic efficacy [28]. Nevertheless, there is limited number of studies focusing on the role of abnormal ERK/MAPK signaling in neuroblastoma cancer. In one study, it was shown that this signaling pathway is responsible for the transformation of neuroblastoma cells and gaining resistance to chemotherapy [29]. In this study, they incubated SKNSH neuroblastoma cell lines with increasing concentrations of doxorubicin or MDL 28842 for long-term to establish drug resistant SKNSH cell lines. Then they analyzed the levels of epidermal growth factor receptor (EGFR) expression and epidermal growth factor (EGF)-induced EGFR tyrosine phosphorylation and determined that they were both lower in drug-resistant SKNSH cells compared with their wild-type counterparts. In addition, in doxorubicin treated SKNSH cells, MAPK activation and nuclear translocation were found to be decreased in response to EGF. These results reveal that chemotherapeutic drug resistance in human neuroblastoma cell lines is in close association with low levels of growth factor signaling through the MAPK pathway.

While continuous activation of the ERK/MAPK signaling pathway promote the transformation of normal cells into tumor cells, inhibition of the ERK/MAPK signaling can restore tumor cells to their non-transformed state in vivo and in vitro [27]. In our laboratory, we examined the effect of ERK/MAPK inhibition on the SH-SY5Y neuroblastoma cell viability. Results of MTS cell viability analysis showed that the viability of SH-SY5Y neuroblastoma cells upon treatment with the specific MEK1/2 inhibitor U0126 was significantly decreased. This result indicates a close link between ERK/MAPK pathway and carcinogenesis in SH-SY5Y neuroblastoma cells [30].

On the other hand, iron chelators have also been used to inhibit the ERK/MAPK signaling pathway and it has been shown in prostate cancer cells that they are able to regulate the ERK/MAPK signaling by reducing ERK1/2 phosphorylation [13, 31]. Based on the results of these studies, we examined the anti-proliferative effects of iron-chelating salicylidene acylhydrazide group synthetic compounds ME0053, ME005 and ME0192 in SH-SY5Y neuroblastoma cells by analyzing the effects of these compounds on ERK/MAPK and PI3K/AKT activities. The results indicated that these iron-chelators caused a significant decrease in MEK1/2 expression and AKT phosphorylation in neuroblastoma cells [24]. These results are promising for alleviation of the ERK/MAPK activation by utilizing different iron chelators to prevent cancer.

Furthermore, current studies have shown that non-steroidal drugs containing salicylic acid (SA) decrease mortality through mitogenic MEK1/2 protein, an important member of ERK/MAPK signaling in many cancers putting a spotlight on SA as a potential inhibitor of MEK1/2 signaling in the prevention of carcinogenic progression [32–34]. In our laboratory, we studied with a salicylic acid analog acibenzolar-S-methyl to analyze its effects on MEK1/2 signaling in SH-SY5Y neuroblastoma cells and we showed that acibenzolar-S-methyl negatively affects MEK1/2 signaling causing apoptotic death of SH-SY5Y neuroblastoma cells [35]. Besides, in one study conducted with A549 human lung cancer cells, SA has been demonstrated as a suppressor of this vital signaling pathway by inhibiting the binding of c-Raf to Ras protein, disrupting phosphorylation state of c-Raf and thereby damaging the MAPK signaling [36]. The great number of protein kinases modulated by salicylate may be explanatory for the question 'what is the apoptotic mechanism of salicylate in cancer?' [36, 37].

Moreover, in an effort to suppress ERK/MAPK signaling in neuroblastoma, Woodfield et al. [38] hypothesized that inhibiting ERK/MAPK signaling through the novel MEK1/2 inhibitor binimetinib may be effective in neuroblastoma models. For this purpose, they analyzed the response of binimetinib-sensitive and binimetinib-resistant neuroblastoma cells from tissue samples and neuroblastoma cell lines by examining total and phosphorylated MEK and ERK levels. They demonstrated that both primary neuroblastoma tumor cells and cell lines showed significant levels of total and phosphorylated MEK and ERK, while binimetinib treatment caused complete loss of phosphorylated ERK. However, resistant cells showed negligible effects on ERK and MEK phosphorylation. They also showed that Ras-GTPase activating protein (RasGAP) NF1 expression was in correlation with responses to binimetinib, suggesting a potential role for NF1 and ERK/MAPK signaling in neuroblastoma differentiation, drug resistance and relapse [38].

Even though it is well known that ERK/MAPK signaling inhibition results in apoptotic death in many cancers, in certain types of cancer such as melanoma, suppressing this signal may inversely contribute to cancer formation by creating an anti-apoptotic effect [37]. This contradiction draws attention to the heterogeneous and unique nature of the molecular basis of cancer emphasizing the vitality of thoroughly understanding of molecular and cellular mechanisms of each cancer type.

3. PI3K/AKT signaling pathway in cancer

Similar to the ERK/MAPK pathway, PI3K/AKT signaling pathway is activated by the interaction of a growth factor with a RTK that regulates basic cellular functions such as growth, proliferation, cellular metabolism, cytoskeletal organization, survival and apoptosis in normal cells [39]. PI3K, is a member of lipid kinase, is divided into three classes: classes I, II, and III according to its specific substrates and structures. The Class I PI3Ks which are composed of p55 and p85 regulatory subunits (p85a, p55a, p50a, p85b, p55g) and p110 catalytic subunit (p110a, p110b, p110d) is the most frequently associated class with cancer [40]. In normal cellular conditions, PI3K is activated by growth factors, cytokines, and hormones. Following this activation, PI3K triggers the phosphorylation reaction of PtdIns (4,5) P2(PIP2) to produce PtdIns (3,4,5) P3(PIP3).

The most important downstream effector protein of PI3K is a serine/threonine kinase AKT/protein kinase B (PKB) that regulates several mechanisms in cell survival and cell cycle progression [41]. In order to activate AKT signaling, the AKT protein is subjected to successive phosphorylation through Thr308 and Ser473 residues. Semi-active form of AKT protein is achieved by Thr308 phosphorylation, while a sequential phosphorylation on Ser473 region at the C-terminal end by PDK2 (phosphoinositide dependent protein kinase 2) leads to full activation of AKT. Activated AKT leaves the membrane and translocates to the cytoplasm and nucleus. Here, by phosphorylating a wide range of target proteins such as MDM2, mTOR, GSK3β and BAD, it causes cellular responses such as cell proliferation, survival, growth, DNA repair and suppression of apoptosis [28].

The negative regulator of the PI3K/AKT signaling pathway is the Phosphatase and Tensin homolog protein (PTEN), which has been defined as a tumor suppressor and is frequently affected by mutations in cancers. PTEN's substrate is PIP3, one of the PI3K products. PTEN inhibits the PI3K/AKT pathway activity by reducing the amount of PIP3, converting PIP3 back to PIP2 via dephosphorylation [42]. PTEN stability and activity is post-translationally regulated by Protein Inhibitor of Activated STAT xα (PIASxα) which is a SUMO E3 ligase for PTEN. PIASxα SUMOylates and stabilizes PTEN protein, thereby negatively regulates PI3K/AKT signaling and leads to G0/G1 cell cycle arrest, and cell proliferation inhibition [43].

Moreover, ubiquitination is another way of post-translational regulation of PI3K/AKT pathway. p85 subunit of PI3K is ubiquitinated and degraded by trosine-phosphorylated c-Cbl E3 ligase which ultimately leads to downregulation of PI3K/AKT signaling [44, 45]. Besides p85 ubiquitination, both caspase- and proteasome-dependent AKT degradation may downregulate PI3K/AKT signaling in case of vascular endothelial growth factor (VEGF) deprivation, mTOR inhibition, or TNF-α treatment [46, 47].

AKT signaling controls metabolic processes either directly, by regulating metabolic enzymes through phosphorylation, or indirectly, by regulating a number of transcription factors. Metabolic enzyme phosphorylation provides short-term changes in the metabolic pathways, while controlling gene expression through the phosphorylation of transcription factors allows for longer-term changes in intracellular metabolic pathways.

Even though AKT is primarily a survival kinase, it also enhances cell proliferation. Cyclin D-1 is a cell cycle regulator which is responsible for G1 to S phase progression. GSK3β phosphorylates cyclin D-1, enabling its transport from nucleus to the cytoplasm for degradation and thereby inhibiting cell cycle. AKT triggers cell proliferation not only by inhibiting this GSK3β kinase activity through phosphorylation, but also by downregulating cyclin dependent kinase inhibitors KIP1 (p27) and CIP1 (p21) [48–50].

Besides cell proliferation, the PI3K/AKT pathway has also been shown to be functional in physiologic and pathologic angiogenesis in animal models [51, 52]. In tumors, PI3K/AKT pathway exerts its pro-angiogenic effects through upregulating HIF-1α, thereby activating VEGF [53]. HIF-1, a heterodimeric protein with α and β subunits, is an activator of VEGF transcription [54]. Moreover, there is data indicating a HIF-1α-independent pathway for PI3K-mediated VEGF upregulation through phosphorylation and activation of endothelial nitric oxide synthase by AKT [55, 56].

Aberrant regulation and activation of the PI3K/AKT pathway is frequent in numerous human malignancies playing a pivotal role in both cancer progression and drug resistance. PI3K/AKT activation is mainly a consequence of the loss of tumor suppressor gene PTEN [57, 58], oncogenic activation of PIK3CA [59, 60] and over-activation by a number of growth factors such as IGF-1, VEGF or EGF [61–63].

Loss-of-function mutations in the PTEN gene are extremely common among sporadic glioblastomas, melanomas, prostate cancers, and endometrial carcinomas. PTEN is negative regulator of the PI3K/AKT signaling pathway that dephosphorylates PIP3. Mutated PTEN leads to increased level of PIP3 that trigger continuous phosphorylation of AKT, thereby leading to continuous activation of the PI3K/AKT signaling pathway. Hyper-activated AKT promotes the survival of cancer cells by causing increased level of cell proliferation and resistance to apoptosis [64, 65].

Although it is obvious that PI3K/AKT also contributes to development of neuroblastoma, its molecular mechanism is poorly understood. Johnsen et al. [66] suggested a link between PI3K/AKT pathway and neuroblastoma through over-activated AKT which appears to be closely related to the disease outcome. In other studies, PI3K/AKT pathway activation was identified as a predictor of poor outcome in neuroblastoma, supporting the afore-mentioned study results and making it a clinically important therapeutic target [67–69]. In one of these studies, they analyzed the effect of small molecule PI3K inhibitors on chemosensitivity in neuroblastoma cell lines and primary cultured neuroblastoma samples. The results of the study showed that PI3K inhibitors, (PI103 for this study), work synergistically with certain chemotherapeutics (Doxorubicin, Etoposide, Topotecan, Cisplatin, Vincristine and Taxol) to drive neuroblastoma cells through apoptosis. PI103 elicits this function by cooperating with chemotherapeutics to decrease the PI3K-mediated inhibitory phosphorylation of pro-apoptotic BimEL, thereby turning the situation in favor of pro-apoptotic Bcl-2 proteins to trigger apoptosis. Thus, targeting PI3K/AKT presents a promising strategy to sensitize neuroblastoma cells for chemotherapy-induced apoptosis [67].

On the other hand, in a study with a murine model of neuroblastoma, they showed that inhibiting PI3K/AKT signaling prevents tumor progression through an effect on oncogenic Mycn protein stability by inactivating GSK3β [70]. Furthermore, AKT phosphorylation has been detected in different neuroblastoma cell lines such as SK-N-SH, SH-SY5Y, SK-N-BE, SH-EP, and IMR-32. Studies related with neuroblastoma cell lines revealed that activated AKT cause poor prognosis and the use of inhibitors specific to the PI3K/AKT signaling pathway leads cancer cells to apoptosis [67, 71]. In another study, SH-SY5Y neuroblastoma cells were exposed to interferon-β resulting in the downregulation of AKT and subsequent apoptosis [72].

4. Pathological interaction of ERK/MAPK and PI3K/AKT signaling pathways in neuroblastoma

Although the usual signaling networks of hormone, cytokine, and growth factor receptors present PI3K/AKT and ERK/MAPK as two independent pathways, there are several inter-pathway cross talk nodes as well as certain regulatory molecules that can simultaneously act on both pathways which together determine the fate of the cell [13, 31]. Based on this information, it can be stated that it is possible for the PI3K/AKT and ERK/MAPK pathways to affect each other either negatively or positively at different signal propagation stages (**Figure 1**).

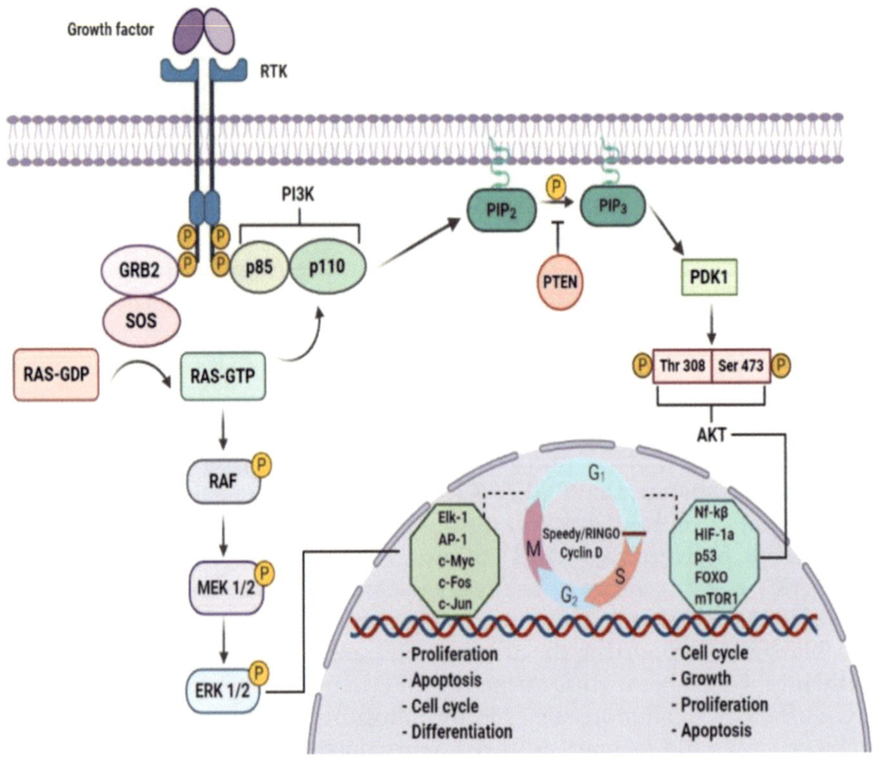

Figure 1.
Interactions of PI3K/AKT and ERK/MAPK pathways.

Examining the mutual talking points of these signaling pathways reveals that the activated RAS protein appears to have a binary switch function triggering both ERK/MAPK and PI3K/AKT pathways [61, 73]. Another important cross talk is the induction of Raf and MEK by PI3K. Cross talk interactions caused by PI3K activation and mediated by PDK1 activate the ERK/MAPK pathway, while AKT and downstream effectors, mTOR and p70S6K, negatively affect ERK/MAPK signaling [28]. On the other hand, active ERK can influence the PI3K/AKT pathway in different interaction routes. One mechanism involves modulation of the ERK-mediated phosphorylation levels at serine and threonine residues of certain AKT members (**Figure 1**) [74]. Moreover, Tuberous Sclerosis Complex 2 (TSC2) can be phosphorylated by either PI3K/AKT pathway or ERK/MAPK pathway, that allows for mTOR activation which is a component of PI3K/AKT signaling [75].

Investigating their interaction in terms of carcinogenesis, it is apparent that their abnormal interaction has an incontrovertible effect particularly on the aggressive progression of different types of cancers. In one of the studies examining their interaction in cancer, the role of the ERK/MAPK pathway in the control of self-renewal and tumorigenicity of glioblastoma cancer stem-like cells (CSLCs) was investigated in relation to the PI3K/AKT pathway. When they inactivated MEK1/2 using chemical inhibitors or siRNA, both cell line- and patient derived glioblastoma CSLCs were shown to lose their spherical form and differentiate into neuronal and glial lineages. Further, this observed effect of MEK inactivation was enhanced by using a dual inhibitor (NVP-BEZ235) of PI3K and mTOR suggesting that inactivating either ERK/MAPK or PI3K/AKT pathway leads to activation of the other, implying the presence of a mutual inhibitory cross talk between them [4].

In another study emphasizing the importance of concomitant inhibition of both pathways in terms of preventing the contributing effect of their cross talk on carcinogenic progression, a mouse model co-expressing the activated forms of AKT and Ras in the liver of mouse was utilized. They showed that continuously and simultaneously activated ERK/MAPK and PI3K/AKT signaling causes accelerated liver tumor development through activation of mTOR. In order to reveal the exact role of mTOR activation in AKT/Ras induced hepatocellular carcinoma, they treated AKT/Ras mice with mTOR inhibitor Rapamycin, and they found out that Rapamycin significantly prevented tumor formation in the liver of AKT/Ras mice. However, Rapamycin withdrawal resulted in rapid recurrence of hepatocellular carcinoma arising from the residual lesions in the liver of AKT/Ras mice through upregulation of ERK and its downstream effectors, Mnk1 and eIF4E in the lesions. Simultaneously suppressing PI3K/AKT and ERK/MAPK pathways was shown to significantly inhibited the growth of AKT/Ras cells in vitro, indicating that there is apparently a sophisticated interaction between these two pathways [5].

As in glioblastoma and hepatocarcinoma, cross talk between the PI3K/AKT and ERK/MAPK pathways contribute to carcinogenesis in rhabdomyosarcoma which is a rare cancer type of soft tissue. Since there is a multifaceted cross talk and a reciprocal compensation between them, blocking both pathways concomitantly in order to have a synergistic inhibitory effect on rhabdomyosarcoma progression was found to be more effective both in vivo and in vitro [6].

Although the effect of their aberrant cross talk is evident in many cancers, there is not sufficient number of studies conducted with neuroblastoma cell lines or primary neuroblastoma cells to investigate their oncogenic interaction in this cancer. On the other hand, ERK/MAPK and PI3K/AKT pathways, in addition to the interactions between their different components, can be affected by other proteins that are not members of these pathways [21]. However, the effect of these interactions on the emergence of cancer, especially of neuroblastoma, is not yet known. At this point, there are data that strengthen the possibility that one of the proteins likely to have an effect on PI3K/AKT and ERK/MAPK signaling pathways is the Speedy/RINGO protein, which is an unconventional cell cycle regulator and plays a very important role in many cancers [76].

Speedy/RINGO binds to its partner Cyclin-dependent kinase 2 (CDK2) and controls G1-S phase transition in the cell cycle [73]. In order to elicit this function, unlike classical cyclins, Speedy/RINGO does not require phosphorylation, and it is also resistant to phosphorylating inhibition by cell cycle inhibitors such as p21Cip1 and p27Kip1 [24]. Because of these properties, Speedy/RINGO can inhibit apoptosis and sustain cancerous cell division by overriding many cell cycle checkpoints [77, 78]. There are various studies conducted with neuroblastoma and breast cancer cells showing the contribution of Speedy/RINGO over-expression together with its partner CDK2 to the carcinogenic process [79, 80].

Apart from these studies, Speedy/RINGO protein has also been shown to have an interaction with ERK/MAPK signaling pathway in a study investigating the tumor formation in breast tissue [81]. In this study, they determined that activating ERK/MAPK pathway resulted in Speedy/RINGO over-expression, and inhibiting this signaling decreased Speedy/RINGO expression. Besides, with another study, it was shown that Speedy/RINGO over-expression leads to the increased activity of its partner proteins, CDK2 and Cyclin A [82]. On the other hand, studies with mouse embryonic stem cell indicated that the Cyclin A2 and CDK2 take part in AKT over-phosphorylation and activation (**Figure 1**) [83].

Considering the results of all these studies has led us to raise the question that "could there be an either direct or indirect three-way interaction between these three players, Speedy/RINGO, ERK/MAPK and PI3K/AKT pathways in neuroblastoma cells?". To analyze this interaction, Speedy/RINGO gene expression was silenced by siRNA in SH-SY5Y neuroblastoma cells in order to determine its effect on the activity of ERK/MAPK and PI3K/AKT pathways. Results showed that silencing Speedy/RINGO in neuroblastoma cells significantly decreased MEK1/2 expression in ERK/MAPK pathway, and AKT Thr308 and Ser473 phosphorylations in PI3K/AKT pathway. Afterwards, ERK/MAPK signaling was blocked by a specific MEK1/2 inhibitor (U0126) in order to examine the effect of ERK/MAPK inhibition on Speedy/RINGO expression and PI3K/AKT signaling activity in SH-SY5Y cells. As a result, inhibiting ERK/MAPK signaling significantly reduced the expression of Speedy/RINGO and its partners CDK2 and Cyclin A as well as AKT phosphorylation suggesting a reciprocal interaction between Speedy/RINGO and ERK/MAPK and PI3K/AKT signaling pathways [30].

As previously mentioned in this chapter, iron homeostasis is in close relation with the regulation of ERK/MAPK and PI3K/AKT signaling activity. There are a growing number of studies demonstrating the strong effect of iron chelators on these two pathways particularly in prostate cancer [13, 31, 84]. Based on the results of these studies, we have investigated the effects of iron-chelating salicylidene acylhydrazide compounds (ME0053, ME0055 and ME0192) on the ERK/MAPK, PI3K/AKT pathways as well as on Speedy/RINGO expression for the reason that it is likely to be one of the effectors of these two pathways [24]. In this study, it was observed that both MEK1/2 activity and AKT phosphorylation on Thr308 and Ser473 sites were decreased together with a significant decrease in Speedy/RINGO expression emphasizing the effect of different metabolic processes such as iron homeostasis on the post-translational regulation of the members of these two pathways as well as on their interaction with other effector molecules such as Speedy/RINGO.

5. Conclusions

ERK/MAPK and PI3K/AKT pathways are very striking in terms of their contribution to carcinogenesis in many cancers. In this chapter, we have summarized the function of abnormal ERK/MAPK and PI3K/AKT signaling and their cross talk in cancer with an emphasis on neuroblastoma, and discussed their provoking action on the onset, progression, and severity of the disease. All afore-mentioned studies in this chapter will pave the way for better understanding of the aberrant post-translational regulation of oncogenic ERK/MAPK and PI3K/AKT pathways with an ultimate effort for fine-tuning of treatment modalities for cancer.

Acknowledgements

Studies mentioned here that were performed in our laboratory was supported by grants to Aysegul Yildiz from Mugla Sitki Kocman University Scientific Research Project Office, Research and Development Projects (Project Numbers: 17/251 and 17/023).

Conflict of interest

Authors declare that there is no conflicts of interest.

Thanks

We sincerely thank Prof. Dr. Uygar Halis Tazebay from Gebze Technical University, Department of Molecular Biology and Genetics and Prof. Dr. Arzu Karabay Korkmaz from Istanbul Technical University, Faculty of Science and Letters, Molecular Biology and Genetics Department for allowing us to use their laboratory infrastructure. We would also thank Assoc. Prof. Emin Ilker Medine from Ege University Institute of Nuclear Sciences for his help about providing SH-SY5Y neuroblastoma cell line, Prof. Dr. Mikael Elofsson from Umeå University for providing all the salicylidene acylhydrazide compounds, and Prof. Dr. Omur BAYSAL from Mugla Sitki Kocman University for his generous gift of acibenzolar-S-methyl (Syngenta).

Author details

Aysegul Yildiz* and Yesim Kaya
Department of Molecular Biology and Genetics, Faculty of Science, Mugla Sitki Kocman University, Mugla, Turkey

*Address all correspondence to: aysegulunal@mu.edu.tr

References

[1] Chrysanthakopoulos NA, Chrysanthakopoulos PA. Molecular Biology and Cellular Signaling Pathways in Glioblastoma. Mathews Journal of Neurology. 2018;3(1):1-13. URL: https://www.mathewsopenaccess.com/full-text/molecular-biology-and-cellular-signaling-pathways-in-glioblastoma

[2] Sharma BS, Prabhakaran V, Desai AP, Bajpai J, Verma RJ, Swain PK. Post-translational Modifications (PTMs), from a Cancer Perspective: An Overview. Oncogen. 2019;2(3). DOI:10.35702/onc.10012

[3] Vogelstein B, Kinzler KW. Cancer genes and the pathways they control. Nat Med. 2004;10(8):789-799. DOI:10.1038/nm1087

[4] Sunayama J, Matsuda K-I, Sato A, et al. Crosstalk between the PI3K/mTOR and MEK/ERK pathways involved in the maintenance of self-renewal and tumorigenicity of glioblastoma stem-like cells. Stem Cells. 2010;28(11):1930-1939. DOI:10.1002/stem.521

[5] Wang C, Cigliano A, Delogu S, et al. Functional crosstalk between AKT/mTOR and Ras/MAPK pathways in hepatocarcinogenesis: Implications for the treatment of human liver cancer. Cell Cycle. 2013;12(13):1999-2010. DOI:10.4161/cc.25099

[6] Jahangiri A, Weiss WA. It takes two to tango: Dual inhibition of PI3K and MAPK in rhabdomyosarcoma. Clin Cancer Res. 2013;19(21):5811-5813. DOI:10.1158/1078-0432.CCR-13-2177

[7] De Preter K, Vandesompele J, Heimann P, et al. Human fetal neuroblast and neuroblastoma transcriptome analysis confirms neuroblast origin and highlights neuroblastoma candidate genes. Genome Biol. 2006;7(9): R84. DOI:10.1186/gb-2006-7-9-r84

[8] Kaneko Y, Kanda N, Maseki N, et al. Different karyotypic patterns in early and advanced stage neuroblastomas. Cancer Res. 1987;47(1):311-318. PMID: 3791218

[9] Tanaka T, Slamon DJ, Shimoda H, et al. Expression of Ha-ras oncogene products in human neuroblastomas and the significant correlation with a patient's prognosis. Cancer Res. 1988;48(4):1030-1034. PMID: 3276397

[10] Widmann C, Gibson S, Jarpe MB, Johnson GL. Mitogen-Activated Protein Kinase: Conservation of a Three-Kinase Module from Yeast to Human. Physiological Reviews. 1999;79(1):143-180. DOI:10.1152/physrev.1999.79.1.143

[11] Wada T, Penninger JM. Mitogen-activated protein kinases in apoptosis regulation. Oncogene. 2004;23(16):2838-2849. DOI:10.1038/sj.onc.1207556

[12] Platanias LC. Map kinase signaling pathways and hematologic malignancies. Blood. 2003;101(12):4667-4679. DOI:10.1182/blood-2002-12-3647

[13] Dhillon AS, Hagan S, Rath O, Kolch W. MAP kinase signalling pathways in cancer. Oncogene. 2007;26(22):3279-3290. DOI:10.1038/sj.onc.1210421

[14] Basso AD, Kirschmeier P, Bishop WR. Lipid posttranslational modifications. Farnesyl transferase inhibitors. J Lipid Res. 2006;47: 15-31. DOI: 10.1194/jlr.R500012-JLR200

[15] Sebti SM, Der CJ. Opinion: searching for the elusive targets of farnesyltransferase inhibitors. Nat Rev Cancer. 2003;3: 945-951. DOI: 10.1038/nrc1234

[16] Kwon OS, Lee H, Kim YJ, Cha HJ, Song NY, Lee MO. ERK

Dephosphorylation through MKP1 Deacetylation by SIRT1 Attenuates RAS-Driven Tumorigenesis. Cancers, 2020; 12(4), 909. https://doi.org/10.3390/cancers12040909

[17] Lu Z, Xu S, Joazeiro C, Cobb MH, Hunter T. The PHD domain of MEKK1 acts as an E3 ubiquitin ligase and mediates ubiquitination and degradation of ERK1/2. Mol. Cell. 2002;9:945-956. DOI: 10.1016/s1097-2765(02)00519-1

[18] Laine A, Ronai Z. Ubiquitin chains in the ladder of MAPK signaling. Sci. STKE. 2005;2005:re5. DOI: 10.1126/stke.2812005re5

[19] Neckers L. Using natural product inhibitors to validate Hsp90 as a molecular target in cancer. Curr Top Med Chem. 2006; 6: 1163-1171. DOI: 10.2174/156802606777811979

[20] Kubota Y, O'Grady P, Saito H, Takekawa M. Oncogenic Ras abrogates MEK SUMOylation that suppresses the ERK pathway and cell transformation. Nat Cell Biol. 2011; 13(3):282-291. doi: 10.1038/ncb2169. Epub 2011 Feb 20. PMID: 21336309

[21] Dang CV. MYC on the Path to Cancer. Cell. 2012;149(1):22-35. DOI:10.1016/j.cell.2012.03.003

[22] Downward J. Targeting RAS signalling pathways in cancer therapy. Nature Reviews Cancer. 2003;3(1):11-22. DOI:10.1038/nrc969

[23] Burotto M, Chiou VL, Lee J-M, Kohn EC. The MAPK pathway across different malignancies: A new perspective. Cancer. 2014;120(22):3446-3456. DOI:10.1002/cncr.28864

[24] Arziman S, Tanriverdi O, Kucukvardar S, Citil N, Yildiz A. Salicylidene acylhydrazides attenuate survival of SH-SY5Y neuroblastoma cells through affecting mitotic regulator Speedy/RINGO and ERK/MAPK–PI3K/AKT signaling. Med Oncol. 2020;37(8):65. DOI:10.1007/s12032-020-01391-9

[25] Denkert C, Schmitt WD, Berger S, et al. Expression of mitogen-activated protein kinase phosphatase-1 (MKP-1) in primary human ovarian carcinoma. Int J Cancer. 2002;102(5):507-513. DOI:10.1002/ijc.10746

[26] Lee SH, Lee JW, Soung YH, et al. Colorectal tumors frequently express phosphorylated mitogen-activated protein kinase. APMIS. 2004;112(4-5):233-238. DOI:10.1111/j.1600-0463.2004.apm11204-0502.x

[27] Sebolt-Leopold JS, Dudley DT, Herrera R, et al. Blockade of the MAP kinase pathway suppresses growth of colon tumors in vivo. Nature Medicine. 1999;5(7):810-816. DOI:10.1038/10533

[28] Aksamitiene E, Kiyatkin A, Kholodenko BN. Cross-talk between mitogenic Ras/MAPK and survival PI3K/Akt pathways: a fine balance. Biochem Soc Trans. 2012;40(1):139-146. DOI:10.1042/BST20110609

[29] Mattingly RR, Milstein ML, Mirkin BL. Down-regulation of growth factor-stimulated MAP kinase signaling in cytotoxic drug-resistant human neuroblastoma cells. Cell Signal. 2001;13(7):499-505. DOI:10.1016/s0898-6568(01)00173-5

[30] Kaya Y, Kucukvardar S, Yildiz A. Speedy/RINGO protein interacts with ERK/MAPK and PI3K/AKT pathways in SH-SY5Y neuroblastoma cells. Mol Cell Biochem. 2020;473(1):133-141. DOI:10.1007/s11010-020-03813-8

[31] Hubert RS, Vivanco I, Chen E, et al. STEAP: A prostate-specific cell-surface antigen highly expressed in human prostate tumors. Proc Natl Acad Sci U

S A. 1999;96(25):14523-14528. PMID: 10588738

[32] Brasky TM, Bonner MR, Moysich KB, et al. Non-steroidal anti-inflammatory drug (NSAID) use and breast cancer risk in the Western New York Exposures and Breast Cancer (WEB) Study. Cancer Causes Control. 2010;21(9):1503-1512. DOI:10.1007/s10552-010-9579-5

[33] Elwood PC, Morgan G, Pickering JE, et al. Aspirin in the Treatment of Cancer: Reductions in Metastatic Spread and in Mortality: A Systematic Review and Meta-Analyses of Published Studies. PLoS One. 2016;11(4): e0152402. DOI:10.1371/journal.pone.0152402

[34] Klessig DF. Newly Identified Targets of Aspirin and Its Primary Metabolite, Salicylic Acid. DNA Cell Biol. 2016;35(4):163-166. DOI:10.1089/dna.2016.3260

[35] Yildiz A. Acibenzolar-S-Methyl Inhibits MEK1/2 Signaling in SH-SY5Y Neuroblastoma Cells. European Journal of Biology. 2019;78(2):83-88. URL: https://dergipark.org.tr/en/pub/iufsjb/638178

[36] Pan M-R, Chang H-C, Hung W-C. Non-steroidal anti-inflammatory drugs suppress the ERK signaling pathway via block of Ras/c-Raf interaction and activation of MAP kinase phosphatases. Cell Signal. 2008;20(6):1134-1141. DOI:10.1016/j.cellsig.2008.02.004

[37] Haydn JM, Hufnagel A, Grimm J, Maurus K, Schartl M, Meierjohann S. The MAPK pathway as an apoptosis enhancer in melanoma. Oncotarget. 2014;5(13):5040-5053. PMID: 24970815

[38] Woodfield SE, Zhang L, Scorsone KA, Liu Y, Zage PE. Binimetinib inhibits MEK and is effective against neuroblastoma tumor cells with low NF1 expression. BMC Cancer. 2016; 16: 172. DOI: https://pubmed.ncbi.nlm.nih.gov/26925841/

[39] Porta C, Paglino C, Mosca A. Targeting PI3K/Akt/mTOR Signaling in Cancer. Front Oncol. 2014; 4:64. DOI:10.3389/fonc.2014.00064

[40] Engelman JA. Targeting PI3K signalling in cancer: opportunities, challenges, and limitations. Nature Reviews Cancer. 2009;9(8):550-562. DOI:10.1038/nrc2664

[41] Fruman DA, Meyers RE, Cantley LC. Phosphoinositide kinases. Annu Rev Biochem. 1998; 67:481-507. DOI:10.1146/annurev.biochem.67.1.481

[42] Osaki M, Oshimura M, Ito H. PI3K-Akt pathway: its functions and alterations in human cancer. Apoptosis. 2004;9(6):667-676. DOI:10.1023/B:APPT.0000045801.15585.dd

[43] Wang W, Chen Y, Wang S, Hu N, Cao Z, Wang W, Tong T and Zhang X. PIASxα ligase enhances SUMO1 modification of PTEN protein as a SUMO E3 ligase. J Biol Chem. 2014; 289: 3217-3230. DOI: 10.1074/jbc.M113.508515

[44] Swaminathan G, Tsygankov AY. The Cbl family proteins: ring leaders in regulation of cell signaling. J. Cell Physiol. 2006;209:21-43. DOI: 10.1002/jcp.20694

[45] Guenou H, Kaabeche K, Dufour C, Miraoui H, Marie PJ. Down-regulation of ubiquitin ligase Cbl induced by twist haploinsufficiency in Saethre-Chotzen syndrome results in increased PI3K/Akt signaling and osteoblast proliferation. Am. J. Pathol. 2006;169:1303-1311. DOI: 10.2353/ajpath.2006.060102

[46] Riesterer O, Zingg D, Hummerjohann J, Bodis S, Pruschy M. Degradation of PKB/Akt protein by inhibition of the VEGF receptor/mTOR pathway in endothelial cells. Oncogene.

2004;23:4624-4635. DOI: 10.1038/sj.onc.1207596

[47] Medina EA, Afsari RR, Ravid T, Castillo SS, Erickson KL, Goldkorn T. Tumor necrosis factor-α decreases Akt protein levels in 3T3-L1 adipocytes via the caspase-dependent ubiquitination of Akt. Endocrinology. 2005;146:2726-2735. DOI: 10.1210/en.2004-1074

[48] Diehl JA, Cheng M, Roussel MF, Sherr CJ. Glycogen synthase kinase-3beta regulates cyclin D1 proteolysis and subcellular localization. Genes Dev. 1998;12(22):3499-3511. DOI:10.1101/gad.12.22.3499

[49] Mirza AM, Gysin S, Malek N, Nakayama K, Roberts JM, McMahon M. Cooperative Regulation of the Cell Division Cycle by the Protein Kinases RAF and AKT. Mol Cell Biol. 2004;24(24):10868-10881. DOI:10.1128/MCB.24.24.10868-10881.2004

[50] Medema RH, Kops GJ, Bos JL, Burgering BM. AFX-like Forkhead transcription factors mediate cell-cycle regulation by Ras and PKB through p27kip1. Nature. 2000;404(6779):782-787. DOI:10.1038/35008115

[51] Jiang BH, Zheng JZ, Aoki M, Vogt PK. Phosphatidylinositol 3-kinase signaling mediates angiogenesis and expression of vascular endothelial growth factor in endothelial cells. Proc Natl Acad Sci U S A. 2000;97(4):1749-1753. DOI:10.1073/pnas.040560897

[52] Graupera M, Guillermet-Guibert J, Foukas LC, et al. Angiogenesis selectively requires the p110α isoform of PI3K to control endothelial cell migration. Nature. 2008;453(7195):662-666. DOI:10.1038/nature06892

[53] Zhong H, Chiles K, Feldser D, et al. Modulation of hypoxia-inducible factor 1alpha expression by the epidermal growth factor/phosphatidylinositol 3-kinase/PTEN/AKT/FRAP pathway in human prostate cancer cells: implications for tumor angiogenesis and therapeutics. Cancer Res. 2000;60(6):1541-1545. PMID: 10749120

[54] Forsythe JA, Jiang BH, Iyer NV, et al. Activation of vascular endothelial growth factor gene transcription by hypoxia-inducible factor 1. Mol Cell Biol. 1996;16(9):4604-4613. DOI:10.1128/mcb.16.9.4604

[55] Babaei S, Teichert-Kuliszewska K, Zhang Q, Jones N, Dumont DJ, Stewart DJ. Angiogenic Actions of Angiopoietin-1 Require Endothelium-Derived Nitric Oxide. Am J Pathol. 2003;162(6):1927-1936. PMID: 12759249

[56] Sadeghi N, Gerber DE. Targeting the PI3K pathway for cancer therapy. Future Med Chem. 2012;4(9):1153-1169. DOI:10.4155/fmc.12.56

[57] Saal LH, Johansson P, Holm K, et al. Poor prognosis in carcinoma is associated with a gene expression signature of aberrant PTEN tumor suppressor pathway activity. Proc Natl Acad Sci U S A. 2007;104(18):7564-7569. DOI:10.1073/pnas.0702507104

[58] Tang J-M, He Q-Y, Guo R-X, Chang X-J. Phosphorylated Akt over-expression and loss of PTEN expression in non-small cell lung cancer confers poor prognosis. Lung Cancer. 2006;51(2):181-191. DOI:10.1016/j.lungcan.2005.10.003

[59] Aleskandarany MA, Rakha EA, Ahmed MAH, et al. PIK3CA expression in invasive breast cancer: a biomarker of poor prognosis. Breast Cancer Res Treat. 2010;122(1):45-53. DOI:10.1007/s10549-009-0508-9

[60] Kato S, Iida S, Higuchi T, et al. PIK3CA mutation is predictive of poor survival in patients with colorectal cancer. Int J Cancer. 2007;121(8):1771-1778. DOI:10.1002/ijc.22890

[61] Chapuis N, Tamburini J, Cornillet-Lefebvre P, et al. Autocrine IGF-1/IGF-1R signaling is responsible for constitutive PI3K/Akt activation in acute myeloid leukemia: therapeutic value of neutralizing anti-IGF-1R antibody. Haematologica. 2010;95(3):415-423. DOI:10.3324/haematol.2009.010785

[62] Muders MH, Zhang H, Wang E, Tindall DJ, Datta K. Vascular endothelial growth factor-C protects prostate cancer cells from oxidative stress by the activation of mammalian target of rapamycin complex-2 and AKT-1. Cancer Res. 2009;69(15):6042-6048. DOI:10.1158/0008-5472.CAN-09-0552

[63] Puri N, Salgia R. Synergism of EGFR and c-Met pathways, cross-talk and inhibition, in non-small cell lung cancer. J Carcinog. 2008; 7:9. DOI:10.4103/1477-3163.44372

[64] Luo J, Manning BD, Cantley LC. Targeting the PI3K-Akt pathway in human cancer: rationale and promise. Cancer Cell. 2003;4(4):257-262. DOI: 10.1016/s1535-6108(03)00248-4

[65] Efron PA, Chen MK, Iyengar M, Dai W, Nagaram A, Beierle EA. Differential response of neuroblastoma cells to TRAIL is independent of PI3K/AKT. J Pediatr Surg. 2006;41(6):1072-1080. DOI:10.1016/j.jpedsurg.2006.02.001

[66] Johnsen JI, Segerström L, Orrego A, et al. Inhibitors of mammalian target of rapamycin downregulate MYCN protein expression and inhibit neuroblastoma growth in vitro and in vivo. Oncogene. 2008;27(20):2910-2922. DOI:10.1038/sj.onc.1210938

[67] Opel D, Poremba C, Simon T, Debatin K-M, Fulda S. Activation of Akt Predicts Poor Outcome in Neuroblastoma. Cancer Res. 2007;67(2):735-745. DOI:10.1158/0008-5472.CAN-06-2201

[68] Fulda S. The PI3K/Akt/mTOR pathway as therapeutic target in neuroblastoma. Curr Cancer Drug Targets. 2009;9(6):729-737. DOI:10.2174/156800909789271521

[69] Bender A, Opel D, Naumann I, et al. PI3K inhibitors prime neuroblastoma cells for chemotherapy by shifting the balance towards pro-apoptotic Bcl-2 proteins and enhanced mitochondrial apoptosis. Oncogene. 2011;30(4):494-503. DOI:10.1038/onc.2010.429

[70] Chesler L, Schlieve C, Goldenberg DD, et al. Inhibition of phosphatidylinositol 3-kinase destabilizes Mycn protein and blocks malignant progression in neuroblastoma. Cancer Res. 2006;66(16):8139-8146. DOI:10.1158/0008-5472.CAN-05-2769

[71] Kim S, Kang J, Qiao J, Thomas RP, Evers BM, Chung DH. Phosphatidylinositol 3-kinase inhibition down-regulates survivin and facilitates TRAIL-mediated apoptosis in neuroblastomas. J Pediatr Surg. 2004;39(4):516-521. DOI:10.1016/j.jpedsurg.2003.12.008

[72] Dedoni S, Olianas MC, Onali P. Interferon-β induces apoptosis in human SH-SY5Y neuroblastoma cells through activation of JAK-STAT signaling and down-regulation of PI3K/Akt pathway. J Neurochem. 2010;115(6):1421-1433. DOI:10.1111/j.1471-4159.2010.07046.x

[73] Cao Z, Liao Q, Su M, Huang K, Jin J, Cao D. AKT and ERK dual inhibitors: The way forward? Cancer Lett. 2019; 459:30-40. DOI:10.1016/j.canlet.2019.05.025

[74] Lehr S, Kotzka J, Avci H, et al. Identification of Major ERK-Related Phosphorylation Sites in Gab1. Biochemistry. 2004;43(38):12133-12140. DOI:10.1021/bi049753e

[75] Ma L, Chen Z, Erdjument-Bromage H, Tempst P, Pandolfi PP. Phosphorylation and functional inactivation of TSC2 by Erk implications for tuberous sclerosis and cancer pathogenesis. Cell. 2005;121(2):179-193. DOI:10.1016/j.cell.2005.02.031

[76] Yildiz A, Tanriverdi O. MAPK and AKT Pathway Intersection in Neuroblastoma Cells. Current Trends in Biomedical Engineering & Biosciences. 2017;2(1):17-20. DOI: 10.19080/CTBEB.2017.02.555580

[77] Lubanska D, Porter LA. The atypical cell cycle regulator Spy1 suppresses differentiation of the neuroblastoma stem cell population. Oncoscience. 2014;1(5):336-348. DOI:10.18632/oncoscience.36

[78] Chauhan S, Zheng X, Tan YY, et al. Evolution of the Cdk-activator Speedy/RINGO in vertebrates. Cell Mol Life Sci. 2012;69(22):3835-3850. DOI:10.1007/s00018-012-1050-1

[79] Matsuo T, Seth P, Thiele CJ. Increased expression of p27Kip1 arrests neuroblastoma cell growth. Med Pediatr Oncol. 2001;36(1):97-99. DOI:10.1002/1096-911X(20010101)36:1<97::AID MPO1022>3.0.CO;2-X

[80] Molenaar JJ, Ebus ME, Geerts D, et al. Inactivation of CDK2 is synthetically lethal to MYCN over-expressing cancer cells. Proc Natl Acad Sci U S A. 2009;106(31):12968-12973. DOI:10.1073/pnas.0901418106

[81] Golipour A, Myers D, Seagroves T, et al. The Spy1/RINGO Family Represents a Novel Mechanism Regulating Mammary Growth and Tumorigenesis. Cancer Res. 2008;68(10):3591-3600. DOI:10.1158/0008-5472.CAN-07-6453

[82] Liu M, Cheng Y, Jia M. LM23 is essential for spermatogenesis in *Rattus norvegicus*. Front Biosci (Elite Ed). 2010; 2:187-194. DOI:10.2741/e81

[83] Liu P, Begley M, Michowski W, et al. Cell-cycle-regulated activation of Akt kinase by phosphorylation at its carboxyl terminus. Nature. 2014;508(7497):541-545. DOI:10.1038/nature13079

[84] Kovacevic Z, Chikhani S, Lui GYL, Sivagurunathan S, Richardson DR. The iron-regulated metastasis suppressor NDRG1 targets NEDD4L, PTEN, and SMAD4 and inhibits the PI3K and Ras signaling pathways. Antioxid Redox Signal. 2013;18(8):874-887. DOI:10.1089/ars.2011.4273

Chapter 3

Epigenetic Control of Mesenchymal Stromal Cell Fate Decision

Haoli Ying, Ruolang Pan and Ye Chen

Abstract

Mesenchymal stem cells (MSCs) are progenitors of connective tissues, which have emerged as important tools for tissue engineering owing to their differentiation potential in various cell types. The therapeutic utility of MSCs hinges upon our understanding of the molecular mechanisms involved in cellular fate decisions. Thus, the elucidation of the regulation of MSC differentiation has attracted increasing attention in recent years. A variety of external cues contribute to the process of MSC differentiation, including chemical, physical, and biological factors. Among the multiple factors that are known to affect cell fate decisions, the epigenetic regulation of MSC differentiation has become a research hotspot. In this chapter, we summarize recent progress in the determination of the effects of epigenetic modification on the multilineage differentiation of MSCs.

Keywords: mesenchymal stromal cells, post-translational modifications, differentiation, cellular fate decision, epigenetic regulation

1. Introduction

Mesenchymal stromal cells (MSCs) are progenitors of connective tissues, initially characterized as plastic adherent, fibroblastic cells, with the potential to differentiate into many types of cells, including predominantly osteoblasts (cells that secrete the matrix of the bones), chondrocytes (cells embedded in the lacunae of the cartilage matrix), and adipocytes (fat-storing cells), under appropriate conditions. MSC studies have progressed rapidly since the initial report of human MSC isolation from bone marrow. MSCs have been shown to reside within the connective tissues of most organs. Owing to their ease of isolation and unique characteristics, MSCs have been widely regarded as potential candidates for tissue engineering and repair. Further, the fate decision of MSCs has also piqued the interest of scientists. During the last two decades, various signaling molecules important to MSC differentiation have been identified, and the epigenetic regulation of MSC differentiation has recently become a research hotspot.

The transformation process of MSCs from a self-renewing state to a specific lineage is always accompanied by changes in cell morphology and function, which are largely determined by the differential expression of genes. Specifically, genes related to self-renewal are turned off, and transcription of cell type-specific genes is activated. Epigenetic regulation refers to the phenotypic change through gene differential expression without DNA sequence alteration, including four main categories: [1] DNA methylation, [2] histone modifications, and [3] chromatin remodeling (nucleosome positioning); and [4] non-coding RNAs. It has been widely reported that epigenetic and post-translational modifications have a broad and far-reaching influence on MSC differentiation at multiple levels. Here, we provide an overview of the recent findings regarding the roles of epigenetic modification in the fate decision of MSCs.

2. DNA methylation

DNA methylation is an important epigenetic modification referring to the addition of a methyl (-CH3) group to the fifth carbon atom of a cytosine ring to form 5-methylcytosine (5-mC). The process is catalyzed by enzymes known as DNA methyltransferases (DNMTs). DNA methylation was the first epigenetic mark to be discovered, and it plays an important role in normal human growth, development, aging, tumorigenesis, and other genetic and epigenetic diseases. This epigenetic mark has the ability to turn genes on or off and can be inherited through cell division. Recent studies have suggested that methylation and demethylation of specific genes, such as *Runx2*, osteopontin (*Opn*), distal-Less homeobox 5 (*Dlx5*), osterix, collagen type 2 (*Col2*), and *Col10*, play key roles in the multi-lineage differentiation of MSCs.

2.1 Osteogenic differentiation of MSCs and DNA methylation

According to numerous studies, DNA methylation is dynamically involved in the osteogenesis of MSCs. Generally, it may be considered that DNA methylation has a repressor role in the promoter regions with CpG islands, blocking gene expression. During osteogenic differentiation, demethylation was observed at specific CpG regions in the promoters of osteogenic lineage-specific genes, such as *RUNX2, DLX5, SP7, SPP1 OPN, COX2,* alkaline phosphatase *(ALP)*, and osteocalcin *(OCN)*, and the expression of these genes was sequentially increased, whereas the expression of pluripotent genes and hypermethylated promoters was downregulated.

Villagra *et al.* observed a significant hypermethylation at the osteocalcin gene locus in undifferentiated MSCs, and the CpG methylation of the osteocalcin promoter significantly decreased upon osteogenesis induction [1]. Dansranjavin *et al.* reported that the differentiation of MSCs was accompanied by a reduced expression of stemness genes such as LIN28, via the hypermethylation of their promoter regions [2]. Arnsdorf *et al.* reported a protocol to promote MSC osteogenic differentiation by applying a mechanical stimulus [3]. According to their results, mechanical stimulation causes the release of Dnmt3b from bone-specific genes, thus leading to promoter demethylation and upregulated gene expression. Yang *et al.* found that the depletion of demethylase Tet1 and Tet2 may hinder the demethylation of the P2rX7 promoter, resulting in a decrease in the osteogenic differentiation capacity of MSCs [4]. Furthermore, the involvement of DNA methylation in the osteogenic differentiation of MSCs has been supported by differentiation studies using demethylating agents. For example, Zhou *et al.* reported that pretreatment with 5-aza-2′-deoxycytidine (5-ADC) drives the osteogenic differentiation of MSCs by enhancing the expression of osteogenic genes (such as *Dlx5*) associated with the demethylation of the CpG shore [5]. Abnormal changes in the methylation modification mechanism in osteogenic differentiation are associated with the occurrence and development of many common skeletal diseases. García-Ibarbia *et al.* analyzed hip fracture samples from patients with osteoporosis and found that the activity of the Wnt signaling pathway in osteoblasts was reduced. The methylation statuses of *Fzd10, Tbl1x, Csnk1e, Wnt8A, Csnk1a1l*, and *Sfrp4* were also observed to be significantly different from those found in normal bone tissues [6]. Another study explored the differences in gene-wide DNA methylation patterns in osteoporosis and osteoarthritis. The results showed that there was a total of 241 Cpg sites with significant differences in the methylation status. Bioinformatic analysis showed that the sites of difference were mostly related to cell osteogenic differentiation and skeletal embryonic development, especially the homeobox family genes [7]. Sun *et al.* reported that abnormal CpG island hypermethylation of the ABCB1 gene promoter was correlated with glucocorticoid (GC)-associated osteonecrosis of the femoral head (ONFH) in patients [8].

2.2 Adipogenic differentiation of MSCs and DNA methylation

Adipogenesis is highly regulated by a sequential cascade of transcriptional events. Key transcriptional factors controlling adipogenesis include several CCAAT/enhancer-binding protein (C/EBP) family members, including C/EBPα, β, and δ, and the nuclear receptor peroxisome proliferator γ (PPARγ). On the other hand, a number of negative transcriptional factors have also been identified, including GATA2/3, chicken ovalbumin upstream promoter transcription factor (COUP-TF), interferon regulatory factors (IRFs), and Wnt family proteins.

Barrand *et al.* showed that the promoters of stemness genes (*OCT4, NANOG*, and *SOX2*) were hypermethylated in adipose-derived MSCs [9]. In addition, the promoters of adipogenic genes, including leptin (LEP), PPARγ2, fatty acid-binding protein 4 (FABP4), and lipoprotein lipase (LPL), are hypomethylated in these undifferentiated MSCs, as revealed by Noer *et al* [10]. It was reported by Fujiki *et al.* and Melzner *et al.* that the promoters of PPARγ2 and LEP were progressively demethylated along with the terminal differentiation of adipocytes [11, 12]. Studies also revealed that the levels of the DNA methyltransferase 1 (*DNMT1*) gene transcript increased at the beginning of adipogenesis and then decreased [13], while the levels of *DNMT3a* and *DNMT3b* transcripts increased during differentiation [14]. Wnt signaling is a key determinant of the fate between adipogenic and osteogenic differentiation. Chen *et al.* proved that the methylation level of the Wnt10a 5-region was markedly reduced in MSCs after 5-Aza-dC treatment, which likely significantly inhibited adipogenesis and promoted osteogenesis [15].

2.3 Chondrogenic differentiation of MSCs and DNA methylation

DNA methylation and demethylation status also influence MSC chondrogenic differentiation. DNA methylation at specific CpGs has been shown to influence genes such as *MMP13, IL1, iNOS*, chondromodulin, collagen 9, and *GDF5* in chondrocytes. Similarly, induction of *COL10A1* expression during chondrogenesis of MSCs is correlated with the demethylation of two CpG sites in the *COL10A1* promoter. In addition, Kim *et al.* also showed an elevated extent of DNA methylation in the *SOX9* promoter in damaged chondrocytes of osteoarthritis (OA) patients compared to the observation in normal chondrocytes [16]. More recently, Barter *et al.* characterized the DNA methylation changes during the chondrogenesis of MSCs using an Infinium 450 K methylation array. A chondrocyte-specific methylation profile was established by comparison with cartilage and non-cartilage tissue methylation profiles, and they also identified significant changes in DNA hypomethylation at many key cartilage gene loci during chondrogenic differentiation, including *COL11A2, SOX9*, and *ACAN* [17]. Further studies analyzing these epigenetic changes during chondrogenesis are needed.

3. Histone epigenetic modification

Histone modification, a common form of epigenetic regulation, refers to post-translational modifications that are added to the N-terminal tail of histones. Histone modification has been shown to play important roles in regulating cell-specific gene expression. So far, more than sixty different residues on core histones (H2A, H2B, H3, H4) with potential to be modified have been reported. These modifications made to histones, including acylation, methylation, phosphorylation, ubiquitination, and sumoylation, can impact gene expression by altering the chromatin structure or recruiting histone modifiers. Histone proteins function to package DNA, which wraps around the eight histones, into chromosomes. In general, it has been well established that histones on the promoter regions of master transcription factors associated with MSC cell fate commitment, such as *RUNX2* and *OSX* in osteogenic differentiation, *PPARG* and *CEBPA* in adipogenic differentiation and *SOX9* in chondrogenic differentiation, are dynamically modified (**Table 1**). In response to appropriate developmental and/or differentiation signals, histone modifications act in diverse biological processes such as transcriptional activation/inactivation, chromosome packaging, and DNA damage/repair. Furthermore, different types of modifications may have synergistic or antagonistic effects to regulate specific gene expression.

3.1 Histone acetylation modification

Histone acetylation is an epigenetic modification characterized by the addition of an acetyl group ($COCH_3$) to histone proteins, specifically to lysine residues within the N-terminal tail. Histone acetylation is one of the most common epigenetic modifications, which leads to the neutralization of the positive charge on the histone proteins, weakening their interaction with DNA, and finally promoting the opening of chromatin structure and activating gene transcription. On the other hand, histone deacetylation is related to chromatin transcription inhibition. The level of histone acetylation is mainly regulated by histone acetylase (HAT) and histone deacetylase (HDAC).

The degree of histone acetylation of related regulatory genes can reflect the maintenance of stemness and the differentiation status of MSCs. During the process of osteogenic differentiation, the expression of osteogenic genes (such as *RUNX2*, *OSX*, and *ALP*) gradually increases in MSCs, while the expression of stemness-related genes (such as *OCT4* and *SOX2*) is significantly decreased. These changes in gene expression were found to be closely related to H3K9Ac and H3K14Ac, which can be used as marks of gene activation. The regulatory roles of histone acetylases in osteoblast differentiation have been increasingly recognized. According to previous and existing studies, *HDAC1, HDAC6, HDAC8*, and *SIRT1* play important roles in the differentiation of MSCs. For example, Wang *et al.* found that inhibiting the expression of histone deacetylase 1 (HDAC1) can effectively enhance the osteogenic differentiation, gene expression, and the bone formation activity of bone marrow MSCs under mechanical stimulation [25]. Lu *et al.* showed that HDAC1 has a negative correlation with cardiac cell differentiation of MSCs under a myocardial microenvironment. During this process, the expression of HDAC1 in MSCs was significantly decreased in a time-dependent manner. In addition, their data proved that the knockdown of HDAC1 promoted the directed differentiation of MSCs into cardiac cells [26]. Several studies have shown that the expression of HDAC6 is reduced during osteogenic differentiation, and HDAC6 negatively regulates the expression of *OC*, osteopontin *(OPN), BSP2, OSX*, and *ALP* partly by binding to the *RUNX2* C-terminus and adjusting *RUNX2* activity [27–29]. Fu *et al.* reported

Involved epigenetic histone modification	Factor	Mechanism	Result	Reference
Histone deacetylation	HDAC inhibitor	Stimulate the transcription of $p21_{CIP1/WAF1}$ through enhancing the H3 and H4 acetylation	Arrest the cell cycle at the G2/M check point, inhibit adipogenic, chondrogenic, and neurogenic differentiation; promote osteogenesis	[18]
Histone acetylation	Knockdown of PCAF (histone H3K9 acetyltransferase)	Insufficient to increase H3K9 acetylation at promoters of BMP2, BMP4, BMPR2B, and Runx2	INHIBIT adipogenic differentiation and promote osteogenic differentiation in MSCs; reduce the bone formation both in vitro and in vivo	[19]
Histone acetylation	GCN5 knockdown	Insufficient to inhibit NF-κB signaling by mediating the proteasomal degradation of p65 (acetyl K310)	Inhibits osteogenic differentiation of MSCs	[20]
Histone deacetylation	SIRT1 knockout	Insufficient to deacetylate β-catenin to promote its accumulation in the nucleus	Reduce differentiation towards osteoblasts, and chondrocytes	[21]
Histone demethylation	Overexpression of KDM5A	Decrease H3K4me3 levels on promoters of Runx2 by demethylating H3K4me3	Inhibit osteogeninsis; lead to osteoporosis	[22]
Histone methylation	G9a inhibitor	Unclear (correlate with PPARγ and C/EBPα expression)	Impair the proliferation but the anti-proliferative effect is not sustained; increase adipogenic potential and decrease osteogenic potential of MSCs	[23]
Histone methylation	Downregulation of BMI1	Insufficient to recruit and stabilize PRC2 which trimethylate H3K27	Cellular senescence	[24]
Histone methylation	Downregulation of EZH2	Insufficient to trimethylate H3K27 as catalytic subunit of PRC2 and keep a high extent of H3K27me3 to suppress $p16^{INK4A}$-induced senescence	Cellular senescence	[24]

Table 1.
Histone modification in MSC differentiation and aging.

that the pharmacological inhibition of HDAC8 by the HDAC inhibitor, valproic acid (VPA), increased the levels of H3K9Ac and significantly enhanced the expression of the osteogenesis-related genes *RUNX2*, osterix, *OCN*, OPN, and *ALP*. Similarly, knockdown of HDAC8 enhanced the osteogenic differentiation of MSCs [30]. Furthermore, SIRT1, an NAD + -dependent deacetylase, also acts as a key regulator of MSC differentiation. The decrease in its activity reduces the expression of the stemness factor Sox2, which leads to the degradation of the self-renewal and differentiation ability of MSCs [31]. It was reported that SIRT1 knock-out MSCs showed reduced differentiation toward osteoblasts and chondrocytes in vitro, but showed no difference in proliferation or apoptosis. Petra *et al.* showed that SIRT1 deacetylates β-catenin to promote its accumulation in the nucleus, leading to the transcription of genes for MSC differentiation [21]. Additionally, epigenetic research has shed light on the effects of histone acetylation on adipogenesis and chondrogenesis. Chemically distinct HDAC inhibitors have been shown to prevent adipocyte differentiation [32]. It has been reported that SIRT2 acts as an important regulator of adipocyte differentiation through the deacetylation of forkhead box protein O1 (FOXO1), peroxisome proliferator-activated receptor gamma (PPARγ), and CCAAT/enhancer binding protein β (C/EBPβ) [33, 34]. Nakade *et al.* showed that Jun dimerization protein 2 (JDP2) plays a key role as a repressor of adipocyte differentiation by regulating the expression of C/EBPδ via the inhibition of histone acetylation [35]. In addition, SIRT1 can also promote the cartilage differentiation process of MSCs by activating the deacetylation of Sox9 and NF-κB [36].

3.2 Histone methylation

Histone methylation is another common post-translational modification by which methyl groups are transferred to the amino acids of histone proteins that make up the nucleosomes. Histone methylation can occur at various sites in histone proteins, primarily on lysine and arginine residues, and it can be governed by multiple positive and negative regulators, even at a single site, to either activate or repress transcription. Histone methylation is regulated by histone methyltransferase (HMT) and histone demethylase (HDM), which can be monomethylated, dimethylated, or trimethylated.

The increase in methylation usually promotes the affinity of histones to DNA and increases the degree of transcriptional inhibition, such as H3K9 methylation and H3K27 methylation. H3K9 dimethylation and trimethylation are typical repressive histone modifications that mediate the formation of heterochromatic regions. It was reported that the knockdown of ESET, a H3K9 methyltransferase, causes an aberrant expression of Runx2 and finally leads to the impairment of osteogenic differentiation and bone defects in mice. On the other hand, the knockdown of EHMT1, a H3K9 specific methyltransferase, resulted in decreased H3K9me2 levels on the promoters of Runx2, thereby upregulating transcription in mouse tissues. With respect to the adipogenic differentiation of MSCs, it was found that the enrichment of H3K9me1 and H3K9me2 on the promoters of C/EBP and PPARγ was negatively associated with adipogenic differentiation. Lowering the H3K9 methylation levels in these regions by either H3K9 demethylase or HMT inhibitors ultimately promoted adipogenic differentiation. In addition, H3K9me3 levels in the promoter region of Sox9, as well as its target genes *Col2a1* and *aggrecan*, were found to be negatively correlated with the chondrogenic differentiation of MSCs. H3K27 methylation is another heterochromatic histone modification associated with transcriptional repression. H3K27me3 on the promoters of Wnt family genes, including *Wnt1, Wnt6, Wnt10a,* and *Wnt10b*, was increased during osteogenesis. Knockdown of the H3K27me3 demethylases such as KDM4B

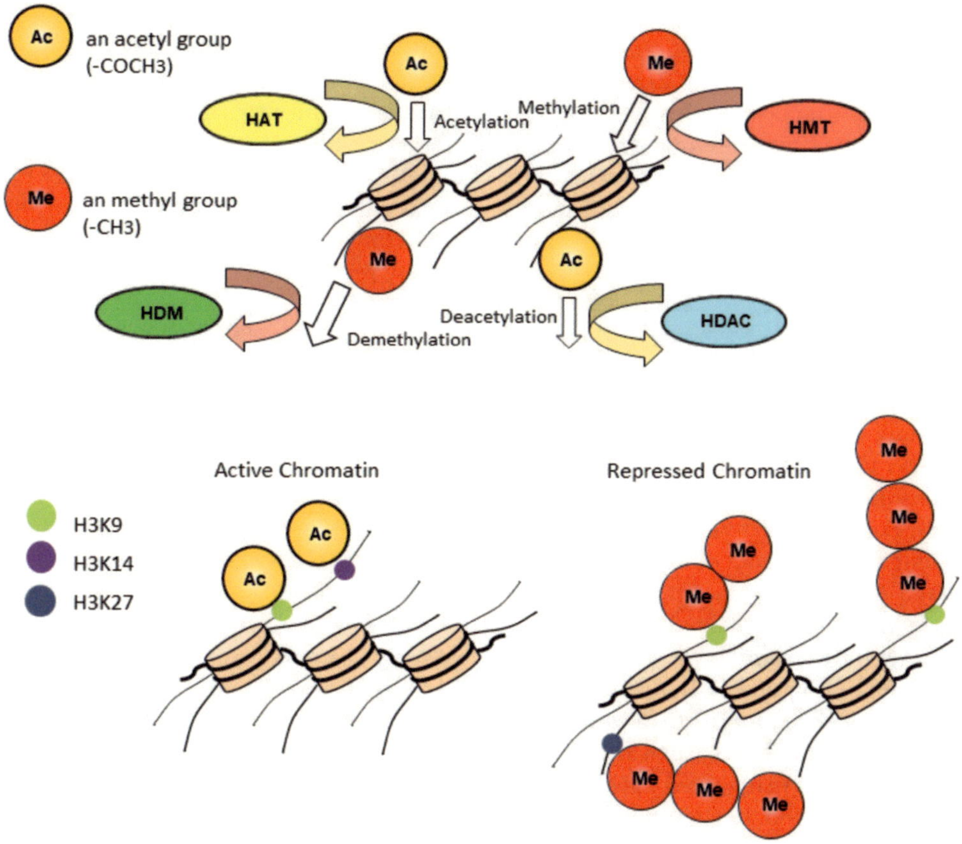

Figure 1.
The mechanisms of histones modification.

and KDM6B attenuated osteogenesis. Moreover, H3K27me3 demethylases have also been reported to facilitate the adipogenic and chondrogenic differentiation of MSCs. Hemming *et al.* found that high expression of the enhancer of zeste homolog 2 (EZH2), a histone methyltransferase catalyzes the tri-methylation of chromatin H3K27, promotes the adipogenic differentiation of MSCs, and inhibits the osteogenic differentiation, while demethylase KDM6A has the opposite effect. In contrast to H3K9 and H3K27 methylation, methylation of H3K4 is associated with an active transcription state. It has been reported that H3K4me3 levels are negatively associated with osteogenesis. HDMs, such as KDM5A and KDM5B, which specifically demethylate H3K4, were found to influence MSC osteogenic differentiation. A recent study also reported that silencing of Ash1l, an H3K4 methyltransferase, promotes adipogenesis while suppressing osteogenesis and chondrogenesis (**Figure 1**).

4. Chromatin remodeling

Chromatin remodeling is the dynamic modification of chromatin architecture, which is an important mechanism for regulating gene expression. In eukaryotes, DNA is tightly wound into a complex called chromatin. Chromatin remodeling allows the access of tightly condensed DNA to various regulatory factors, such as transcription factors and components of DNA replication, so that specific genes can be expressed. The basic mechanism of chromatin remodeling depends on the three dynamic properties of nucleosomes: reconstruction, enzyme-induced covalent modification, and repositioning. In addition, the aforementioned histone modification is another important aspect of chromatin remodeling. Aberrations in chromatin remodeling proteins are associated with various human disorders and diseases. The major activities involved in nucleosome structure alterations use the energy supplied by ATP hydrolysis to affect nucleosomes. These enzymes are called ATP-dependent chromatin (or nucleosome) remodeling factors. The system involves four subfamilies of ATP-dependent chromatin remodeling complexes, namely switch/sucrose non-fermentable (SWI/SNF), nucleosome remodeler deacetylase (NuRD), INO80, and imitation switch (ISWI).

Several studies have demonstrated that functional SWI/SNF machinery plays an important role in regulating MSC tri-lineage differentiation by interacting with tissue-specific transcription factors and crosstalk with cell signaling pathways. Brahma-associated factor (BAF) complex subunits have been implicated in MSC osteo-lineage commitment. For example, depletion of BRG1 leads to constitutive osteo-lineage gene expression [37]. BRM negatively regulates osteocalcin expression [38]. Loss of the classical BAF restricted subunit Pbrm1/Arid2/Brd7 leads to reduced osteogenesis without compromising adipogenesis [39]. It has also been reported that SWI/SNF-dependent chromatin remodeling is involved in MSC adipogenic differentiation. BRG1 overexpression was associated with promoted adipogenic differentiation, which was associated with a marked increase in the differentiation markers PPARγ and LPL [40]. BAF45A was identified as an important regulator of adipogenic differentiation in human MSCs [41]. In addition, other ATP-dependent chromatin remodelers, such as chromodomain helicase DNA binding (CHD) proteins, are also involved in MSC lineage commitment. CHD4 was reported to be implicated in chondrogenesis. Simon *et al.* reported that CHD1 is required for the induction of osteoblast-specific gene expression, extracellular-matrix mineralization, and ectopic bone formation in vivo [42]. CReMM plays a role in mediating the transcriptional response to hormones that coordinate osteoblast function [42, 43]. It was proved by Kumar *et al.* that metastasis-associated gene 1(MTA1) negatively regulates osteo-lineage gene expression [44]. Together, chromatin remodeling plays an important role in MSC lineage commitment.

5. Non-coding RNAs

The RNA world is divided into two classes: 1) RNAs that have coding potential (mRNAs) and 2) RNAs without coding potential, referred to as non-coding RNAs (ncRNAs). Although mRNAs have been extensively studied, ncRNAs span more than 98% of DNA transcripts. In the past, these molecules were considered as "evolutionary junk" but increasing evidence suggests that these molecules spatiotemporally regulate protein-coding gene expression in several molecular mechanisms. With improved RNA-sequencing techniques, in recent years, there have been great advances in identifying and understanding ncRNAs. Epigenetic ncRNAs, including microRNAs (miRNAs), small interfering RNAs (siRNAs), piwi-interacting RNA (piRNA), and long noncoding RNAs (lncRNA), have been reported to play key roles in the regulation of various diseases and biological processes, including cellular differentiation, proliferation, apoptosis, gene regulation, and cancer development.

5.1 Long non-coding RNA

lncRNA is a novel class of noncoding RNAs longer than 200 nt, which can regulate gene expression at the transcriptional and post-transcriptional levels. LncRNAs are mainly located in the cell nucleus or cytoplasm, affecting the status and fate of cells through different post-transcriptional mechanisms. Nuclear lncRNAs guide chromatin modifiers, such as DNA methyltransferase, histone methyltransferase, and heteronuclear ribosome protein, to a specific genetic locus and induce chromatin structure remodeling, which in turn regulates gene expression either positively or negatively. Cytoplasmic lncRNAs can either block the functional site or alter the structure and modification of specific proteins, thereby regulating the function and stabilization of these proteins, and ultimately alter the fate and function of cells. During the last decade, multiple studies have demonstrated that lncRNAs are widely involved in growth and development by controlling the fate of cells, including MSCs.

Studies have demonstrated the importance of lncRNAs in bone regeneration and bone formation. Many lncRNAs regulating the osteogenic differentiation of MSCs have been identified, including ANCR, AK141205, AK028326, DANCR, MALAT1, MEG3, MORD, and POIR; these either promote or inhibit osteogenic differentiation through diverse pathways. For example, MALAT1 promotes *OSX* expression and the osteogenesis of MSCs by sponging miR-143, and MALAT1 can be used as a biomarker for the detection of osteoporosis [45]. MEG3 inhibited bone morphogenetic protein 2 (BMP2) through interaction with hnRNPI, which plays an active role in mRNA splicing, and finally suppresses osteogenic differentiation [46]. Exosome-transferred lncRNA, RUNX2 antisense RNA 1 (RUNX2-AS1), decreases the expression of RUNX2 in MSCs by forming an RNA duplex with RUNX2, consequently suppressing osteogenesis [47]. MEG3 has also been shown to promote osteogenic differentiation in MSCs via the BMP4 signaling pathway [48]. Moreover, lncRNAs are associated with osteogenesis through the regulation of classical signaling pathways, including the Wnt/β-catenin pathway, p38 mitogen-activated protein kinase (MAPK) pathway, Notch signaling pathway, and nuclear factor-κB (NFκB) signaling pathway. Currently, few studies have focused on lncRNA expression and their functions in the chondrogenic and adipogenic differentiation of MSCs. Xiao *et al.* reported that adipogenic differentiation-induced noncoding RNA (ADINR) was significantly upregulated in MSCs after adipogenic induction. Knocking out ADINR significantly inhibited the ability of MSCs to differentiate into adipocytes. A mechanistic study revealed that ADINR positively regulates the expression of the transcription factor C/EBPα [49]. Wang *et al.* reported that the expression of the lncRNAs ZBED3-AS1 and CTA-941F9.9 was significantly upregulated during the differentiation process of MSCs toward cartilage [50].

5.2 MicroRNA

MicroRNAs are the most abundant class of small ncRNAs with a length of 21–25 nt, and have been studied extensively. miRNAs are also involved in the epigenetic regulation of genes in both the cytoplasm and nucleus through different mechanisms. Their main action is the negative regulation of gene expression by specifically binding to a target mRNA through base complementary pairing and inducing its degradation or the inhibition of its translation.

Accumulating evidence indicates that miRNAs play an important role in the maintenance of stemness and differentiation of MSCs (**Table 2**). As mentioned above, lineage differentiation of MSCs is a complex biological process. For example, MSCs differentiate into osteogenic progenitor cells and subsequently osteoblasts, and then gradually become mature bone cells along with a variety of extracellular matrix mineralization. This process involves a large number of secretory and transcription factors. In addition, the differentiation and maturation of MSCs also involves signaling pathways such as WNT, BMP, and PI3K/Akt. The key effector

Involved miRNA	Mechanism	Result	Reference
miR-23a	targets LRP5 and subsequently suppress the Wnt/β-catenin signaling pathway	Inhibit osteogenesis of MSCs	[51]
miR-26a	in BMSCs: targets GSK3*β* mainly and activates Wnt/β-catenin signaling pathway; in ADSCs: targets Smad1 mainly and inhibits BMP signaling pathway	Inhibit osteogenesis of ADSCs and promote osteogenesis of BMSCs	[52]
miR-30c	reduces Runx2 protein	Inhibit osteogenesis of MSCs	[53]
miR-34c			
miR-133a			
miR-135a			
miR-137			
miR-204			
miR-205			
miR-217			
miR-338			
miR-20b	Activate the BMPs/Runx2 signaling pathway at four levels, which consists of repressing PPAR$_{\gamma}$, Bambi and Crim1	Promote ostegenesis	[54, 55]
miR-29b	activates the AKT/β-catenin signaling pathway by inhibiting PTEN expression	Promote osteogenesis of hADSCs	[56]
miR-196a	targets HOXC8 (a negative regulator of SMAD1)	Inhibit proliferation and promote osteogenesis of hDASCs	[57]
miR-17-5p	Represses the Wnt signaling pathway effector Tcf7l2	Promote adipogenesis of BM-MSCs	[58, 59]
miR-21	Alters SMAD3 phosphorylation without affecting total levels of SMAD3 protein and modulate TGF-β signaling pathway		[59, 60]
miR-143	Directly represses MAP2K5 (a key member of the MAPKK family in the MAPK signaling pathway)		[59, 61]
miR-30a	Targets Runx2	Promote adipogenesis	[62]
miR-30d			
miR-642a-3p	unknown	In a high level in adipogenesis	[62]

Table 2.
miRNA and MSCs differentiation.

molecules in these pathways can be regulated by miRNAs, which in turn affects MSC fate decisions. Recently, various miRNAs, including miR-20b, -29b, -30a-5p, -142-3p, -196a, -210, -746-5p, -2861, -3960, -335-5p, etc., have been reported to enhance osteogenic differentiation, whereas miR -23a, -26a, -30c, -34b, -34c, -125, -133a, -135a, -137, -141, -148, -200a, -204, -205, -206, -217, and -338 could impede osteogenic differentiation, and miR-143, -24, -31, -30c, and -642a-3p are involved in regulating adipogenesis. Oskowitz *et al.* reported that silencing of Dicer or Drosha, two key enzymes in the miRNA biogenesis pathway, inhibits both the osteogenic and adipogenic differentiation of MSCs [63]. Some miRNAs have been reported to act as switches for MSCs to differentiate into different lineages. For example, the miR-17 cluster of the miRNA family, miR-17-5p, miR-106a, and miR-20a, are downregulated when the cell undergoes osteogenic differentiation and is upregulated during adipocyte differentiation [64]. Miyaki et al. also demonstrated that the expression of miR-140 increased during chondrocytic differentiation along with the expression of Sox9, Aggercan, and Col2A1 [65]. In addition, recent research has found that miRNAs can form a competitive endogenous RNA regulation network with lncRNAs and circRNAs. Some research groups have started paying more attention to this regulatory network, which will further improve our understanding of the role of ncRNAs in MSC maintenance and differentiation.

6. Conclusions

"Epigenetics" was first used to define the complex interactions between the genome and the environment that are involved in the development and differentiation of organisms. Nowadays, the term refers to heritable alterations in gene expression that are not mediated at the DNA sequence level. Accumulating evidence has suggested that the processes of epigenetic modifications are crucial and largely responsible for the variable activation and repression of specific genes at specific time points during the lifespan of stem cells, allowing for the terminally differentiated phenotype. With the advances in biological and experimental technologies, a variety of epigenetic modifications involved in the cell fate determination of MSCs have been discovered in recent years. In addition to the types of epigenetic modifications introduced in the article, some researchers have suggested the role of histone phosphorylation, ubiquitination, and other modifications in the differentiation of MSCs. On this basis this information, drugs that effectively regulate these modifications have been developed to provide precise differentiation conditions for MSCs and make them more effective in clinical treatment. The disadvantage of epigenetic therapy using small molecule drugs is the lack of specificity, which needs to be further studied. In summary, epigenetic modifications play an important regulatory role in the cell fate determination of MSCs, but the precise function of these modifications in different MSC types, as well as the associated underlying mechanisms, remain to be thoroughly investigated. In-depth research in this field would provide important reference data for the differentiation mechanism research and clinical application of MSCs.

Acknowledgements

The authors were supported by the National Key Research and Development Program of China (2020YFA0113003, 2018YFC1004803) and the Fundamental Research Funds for the Central Universities.

Conflict of interest

The authors declare no competing financial interests.

Author details

Haoli Ying[1,2], Ruolang Pan[3] and Ye Chen[1,2]*

1 Institute of Genetics and Division of Medical Genetics and Genomics, The Children's Hospital, Zhejiang University School of Medicine, Hangzhou, Zhejiang, China

2 Zhejiang Provincial Key Laboratory of Genetic and Developmental Disorders, Hangzhou, Zhejiang, China

3 Key Laboratory of Cell-Based Drug and Applied Technology Development in Zhejiang Province, Hangzhou, Zhejiang, China

*Address all correspondence to: yechency@zju.edu.cn

References

[1] Villagra A, Gutirrez J, Paredes R, Sierra J, Puchi M, Imschenetzky M, et al. Reduced CpG methylation is associated with transcriptional activation of the bone-specific rat osteocalcin gene in osteoblasts. J Cell Biochem. 2002;

[2] Dansranjavin T, Krehl S, Mueller T, Mueller LP, Schmoll HJ, Dammann RH. The role of promoter CpG methylation in the epigenetic control of stem cell related genes during differentiation. Cell Cycle. 2009;

[3] Arnsdorf EJ, Tummala P, Kwon RY, Jacobs CR. Mechanically induced osteogenic differentiation - The role of RhoA, ROCKII and cytoskeletal dynamics. J Cell Sci. 2009;

[4] Yang R, Yu T, Kou X, Gao X, Chen C, Liu D, et al. Tet1 and Tet2 maintain mesenchymal stem cell homeostasis via demethylation of the P2rX7 promoter. Nat Commun [Internet]. 2018;9(1). Available from: http://dx.doi.org/10.1038/s41467-018-04464-6

[5] Zhou GS, Zhang XL, Wu JP, Zhang RP, Xiang LX, Dai LC, et al. 5-Azacytidine facilitates osteogenic gene expression and differentiation of mesenchymal stem cells by alteration in DNA methylation. Cytotechnology. 2009;

[6] García-Ibarbia C, Delgado-Calle J, Casafont I, Velasco J, Arozamena J, Pérez-Núñez MI, et al. Contribution of genetic and epigenetic mechanisms to Wnt pathway activity in prevalent skeletal disorders. Gene. 2013;

[7] Delgado-Calle J, Fernández AF, Sainz J, Zarrabeitia MT, Sañudo C, García-Renedo R, et al. Genome-wide profiling of bone reveals differentially methylated regions in osteoporosis and osteoarthritis. Arthritis Rheum. 2013;

[8] Sun Z, Yang S, Ye S, Zhang Y, Xu W, Zhang B, et al. Aberrant CpG Islands' hypermethylation of ABCB1 in mesenchymal stem cells of patients with steroid-associated osteonecrosis. J Rheumatol. 2013;40(11):1913-20.

[9] Barrand S, Collas P. Chromatin states of core pluripotency-associated genes in pluripotent, multipotent and differentiated cells. Biochem Biophys Res Commun. 2010;

[10] Noer A, Serensen AL, Boquest AC, Collas P. Stable CpG hypomethylation of adipogenic promoters in freshly isolated, cultured, and differentiated mesenchymal stem cells from adipose tissue. Mol Biol Cell. 2006;

[11] Fujiki K, Kano F, Shiota K, Murata M. Expression of the peroxisome proliferator activated receptor γ gene is repressed by DNA methylation in visceral adipose tissue of mouse models of diabetes. BMC Biol. 2009;

[12] Melzner I, Scott V, Dorsch K, Fischer P, Wabitsch M, Brüderlein S, et al. Leptin gene expression in human preadipocytes is switched on by maturation-induced demethylation of distinct CpGs in its proximal promoter. J Biol Chem. 2002;

[13] Stachecka J, Lemanska W, Noak M, Szczerbal I. Expression of key genes involved in DNA methylation during in vitro differentiation of porcine mesenchymal stem cells (MSCs) into adipocytes. Biochem Biophys Res Commun. 2020;

[14] Nomura Y, Hara ES, Yoshioka Y, Nguyen HT, Nosho S, Komori T, et al. DNA Methylation-Based Regulation of Human Bone Marrow-Derived Mesenchymal Stem/Progenitor Cell Chondrogenic Differentiation. Cells Tissues Organs. 2020;207(3-4):115-26.

[15] Chen YS, Wu R, Yang X, Kou S, MacDougald OA, Yu L, et al. Inhibiting

DNA methylation switches adipogenesis to osteoblastogenesis by activating Wnt10a. Sci Rep. 2016;

[16] Kim K Il, Park YS, Im G Il. Changes in the epigenetic status of the SOX-9 promoter in human osteoarthritic cartilage. J Bone Miner Res. 2013;

[17] Barter MJ, Bui C, Cheung K, Falk J, Gómez R, Skelton AJ, et al. DNA hypomethylation during MSC chondrogenesis occurs predominantly at enhancer regions. Sci Rep. 2020;

[18] Lee S, Park JR, Seo MS, Roh KH, Park SB, Hwang JW, et al. Histone deacetylase inhibitors decrease proliferation potential and multilineage differentiation capability of human mesenchymal stem cells. Cell Prolif. 2009;

[19] Zhang P, Liu Y, Jin C, Zhang M, Lv L, Zhang X, et al. Histone H3K9 Acetyltransferase PCAF Is Essential for Osteogenic Differentiation Through Bone Morphogenetic Protein Signaling and May Be Involved in Osteoporosis. Stem Cells. 2016;

[20] Zhang P, Liu Y, Jin C, Zhang M, Tang F, Zhou Y. Histone Acetyltransferase GCN5 Regulates Osteogenic Differentiation of Mesenchymal Stem Cells by Inhibiting NF-κB. J Bone Miner Res. 2016;

[21] Simic P, Zainabadi K, Bell E, Sykes DB, Saez B, Lotinun S, et al. SIRT1 regulates differentiation of mesenchymal stem cells by deacetylating β-catenin. EMBO Mol Med. 2013;

[22] Wang C, Wang J, Li J, Hu G, Shan S, Li Q, et al. KDM5A controls bone morphogenic protein 2-induced osteogenic differentiation of bone mesenchymal stem cells during osteoporosis. Cell Death Dis. 2016;

[23] Khanban H, Fattahi E, Talkhabi M. In vivo administration of G9A inhibitor A366 decreases osteogenic potential of bone marrow-derived mesenchymal stem cells. EXCLI J. 2019;

[24] Jung JW, Lee S, Seo MS, Park SB, Kurtz A, Kang SK, et al. Histone deacetylase controls adult stem cell aging by balancing the expression of polycomb genes and jumonji domain containing 3. Cell Mol Life Sci. 2010;

[25] Wang J, Wang CD, Zhang N, Tong WX, Zhang YF, Shan SZ, et al. Mechanical stimulation orchestrates the osteogenic differentiation of human bone marrow stromal cells by regulating HDAC1. Cell Death Dis. 2016;

[26] Lu DF, Yao Y, Su ZZ, Zeng ZH, Xing XW, He ZY, et al. Downregulation of HDAC1 is involved in the cardiomyocyte differentiation from mesenchymal stem cells in a myocardial microenvironment. PLoS One. 2014;

[27] Yan GQ, Wang X, Yang F, Yang ML, Zhang GR, Wang GK, et al. MicroRNA-22 Promoted Osteogenic Differentiation of Human Periodontal Ligament Stem Cells by Targeting HDAC6. J Cell Biochem. 2017;

[28] Manzotti G, Torricelli F, Donati B, Sancisi V, Gugnoni M, Ciarrocchi A. HDACs control RUNX2 expression in cancer cells through redundant and cell context-dependent mechanisms. J Exp Clin Cancer Res. 2019;

[29] Ozaki T, Wu D, Sugimoto H, Nagase H, Nakagawara A. Runt-related transcription factor 2 (RUNX2) inhibits p53-dependent apoptosis through the collaboration with HDAC6 in response to DNA damage. Cell Death Dis. 2013;

[30] Fu Y, Zhang P, Ge J, Cheng J, Dong W, Yuan H, et al. Histone deacetylase 8 suppresses osteogenic differentiation of bone marrow stromal cells by inhibiting histone H3K9 acetylation and RUNX2 activity. Int J Biochem Cell Biol. 2014;

[31] Yoon DS, Choi Y, Jang Y, Lee M, Choi WJ, Kim SH, et al. SIRT1 directly regulates SOX2 to maintain self-renewal and multipotency in bone marrow-derived mesenchymal stem cells. Stem Cells. 2014;

[32] Catalioto RM, Maggi CA, Giuliani S. Chemically distinct HDAC inhibitors prevent adipose conversion of subcutaneous human white preadipocytes at an early stage of the differentiation program. Exp Cell Res. 2009;

[33] Wang F, Tong Q. SIRT2 suppresses adipocyte differentiation by deacetylating FOXO1 and enhancing FOXO1's repressive interaction with PPARγ. Mol Biol Cell. 2009;

[34] Jing E, Gesta S, Kahn CR. SIRT2 Regulates Adipocyte Differentiation through FoxO1 Acetylation/Deacetylation. Cell Metab. 2007;

[35] Nakade K, Pan J, Yoshiki A, Ugai H, Kimura M, Liu B, et al. JDP2 suppresses adipocyte differentiation by regulating histone acetylation. Cell Death Differ. 2007;

[36] Choi SM, Lee KM, Ryu SB, Park YJ, Hwang YG, Baek D, et al. Enhanced articular cartilage regeneration with SIRT1-activated MSCs using gelatin-based hydrogel. Cell Death Dis. 2018;

[37] Nguyen KH, Xu F, Flowers S, Williams EAJ, Fritton JC, Moran E. SWI/SNF-Mediated Lineage Determination in Mesenchymal Stem Cells Confers Resistance to Osteoporosis. Stem Cells. 2015;

[38] Flowers S, Nagl NG, Beck GR, Moran E. Antagonistic roles for BRM and BRG1 SWI/SNF complexes in differentiation. J Biol Chem. 2009;

[39] Sinha S, Biswas M, Chatterjee SS, Kumar S, Sengupta A. Pbrm1 Steers Mesenchymal Stromal Cell Osteolineage Differentiation by Integrating PBAF-Dependent Chromatin Remodeling and BMP/TGF-β Signaling. Cell Rep. 2020;

[40] Napolitano MA, Cipollaro M, Cascino A, Melone MAB, Giordano A, Galderisi U. Brg1 chromatin remodelling factor is involved in cell growth arrest, apoptosis and senescence of rat mesenchymal stem cells. J Cell Sci. 2007;

[41] van de Peppel J, Strini T, Tilburg J, Westerhoff H, van Wijnen AJ, van Leeuwen JP. Identification of Three Early Phases of Cell-Fate Determination during Osteogenic and Adipogenic Differentiation by Transcription Factor Dynamics. Stem Cell Reports. 2017;

[42] Baumgart SJ, Najafova Z, Hossan T, Xie W, Nagarajan S, Kari V, et al. CHD1 regulates cell fate determination by activation of differentiation-induced genes. Nucleic Acids Res. 2017;

[43] Marom R, Shur I, Hager GL, Benayahu D. Expression and regulation of CReMM, a chromodomain helicase-DNA-binding (CHD), in marrow stroma derived osteoprogenitors. J Cell Physiol. 2006;

[44] Kumar A, Salimath BP, Schieker M, Stark GB, Finkenzeller G. Inhibition of metastasis-associated gene 1 expression affects proliferation and osteogenic differentiation of immortalized human mesenchymal stem cells. Cell Prolif. 2011;

[45] Gao Y, Xiao F, Wang C, Wang C, Cui P, Zhang X, et al. Long noncoding RNA MALAT1 promotes osterix expression to regulate osteogenic differentiation by targeting miRNA-143 in human bone marrow-derived mesenchymal stem cells. J Cell Biochem. 2018;

[46] Liu Y, Zeng X, Miao J, Liu C, Wei F, Liu D, et al. Upregulation of long noncoding RNA MEG3 inhibits the osteogenic differentiation of

periodontal ligament cells. J Cell Physiol. 2019;

[47] Li B, Xu H, Han H, Song S, Zhang X, Ouyang L, et al. Exosome-mediated transfer of lncRUNX2-AS1 from multiple myeloma cells to MSCs contributes to osteogenesis. Oncogene. 2018;

[48] Chen S, Jia L, Zhang S, Zheng Y, Zhou Y. DEPTOR regulates osteogenic differentiation via inhibiting MEG3-mediated activation of BMP4 signaling and is involved in osteoporosis. Stem Cell Res Ther. 2018;

[49] Xiao T, Liu L, Li H, Sun Y, Luo H, Li T, et al. Long noncoding RNA ADINR regulates adipogenesis by transcriptionally activating C/EBPα. Stem Cell Reports. 2015;

[50] Wang L, Li Z, Li Z, Yu B, Wang Y. Long noncoding RNAs expression signatures in chondrogenic differentiation of human bone marrow mesenchymal stem cells. Biochem Biophys Res Commun. 2015;

[51] Li T, Li H, Wang Y, Li T, Fan J, Xiao K, et al. MicroRNA-23a inhibits osteogenic differentiation of human bone marrow-derived mesenchymal stem cells by targeting LRP5. Int J Biochem Cell Biol. 2016;

[52] Su X, Liao L, Shuai Y, Jing H, Liu S, Zhou H, et al. MiR-26a functions oppositely in osteogenic differentiation of BMSCs and ADSCs depending on distinct activation and roles of Wnt and BMP signaling pathway. Cell Death Dis. 2015;

[53] Zhang Y, Xie RL, Croce CM, Stein JL, Lian JB, Van Wijnen AJ, et al. A program of microRNAs controls osteogenic lineage progression by targeting transcription factor Runx2. Proc Natl Acad Sci U S A. 2011;

[54] Zhang ZJ, Zhang H, Kang Y, Sheng PY, Ma YC, Yang ZB, et al. MiRNA expression profile during osteogenic differentiation of human adipose-derived stem cells. J Cell Biochem. 2012;

[55] He J, Zhang J fang, Yi C, Lv Q, Xie W dong, Li J na, et al. miRNA-mediated functional changes through co-regulating function related genes. PLoS One. 2010;

[56] Xia T, Dong S, Tian J. miR-29b promotes the osteogenic differentiation of mesenchymal stem cells derived from human adipose tissue via the PTEN/AKT/β-catenin signaling pathway. Int J Mol Med. 2020;

[57] Kim YJ, Bae SW, Yu SS, Bae YC, Jung JS. miR-196a regulates proliferation and osteogenic differentiation in mesenchymal stem cells derived from human adipose tissue. J Bone Miner Res. 2009;

[58] Tian L, Song Z, Shao W, Du WW, Zhao LR, Zeng K, et al. Curcumin represses mouse 3T3-L1 cell adipogenic differentiation via inhibiting miR-17-5p and stimulating the Wnt signalling pathway effector Tcf7l2. Cell Death Dis. 2017;

[59] An X, Ma K, Zhang Z, Zhao T, Zhang X, Tang B, et al. MiR-17, miR-21, and miR-143 Enhance Adipogenic Differentiation from Porcine Bone Marrow-Derived Mesenchymal Stem Cells. DNA Cell Biol. 2016;

[60] Jeong Kim Y, Jin Hwang S, Chan Bae Y, Sup Jung J. MiR-21 regulates adipogenic differentiation through the modulation of TGF-β signaling in mesenchymal stem cells derived from human adipose tissue. Stem Cells. 2009;

[61] Chen L, Hou J, Ye L, Chen Y, Cui J, Tian W, et al. MicroRNA-143 regulates adipogenesis by modulating the map2k5-erk5 signaling. Sci Rep. 2014;

[62] Zaragosi LE, Wdziekonski B, Brigand KL, Villageois P,

Mari B, Waldmann R, et al. Small RNA sequencing reveals miR-642a-3p as a novel adipocyte-specific microRNA and miR-30 as a key regulator of human adipogenesis. Genome Biol. 2011;

[63] Oskowitz AZ, Lu J, Penfornis P, Ylostalo J, McBride J, Flemington EK, et al. Human multipotent stromal cells from bone marrow and microRNA: Regulation of differentiation and leukemia inhibitory factor expression. Proc Natl Acad Sci U S A. 2008;

[64] Li H, Li T, Wang S, Wei J, Fan J, Li J, et al. MiR-17-5p and miR-106a are involved in the balance between osteogenic and adipogenic differentiation of adipose-derived mesenchymal stem cells. Stem Cell Res. 2013;

[65] Miyaki S, Nakasa T, Otsuki S, Grogan SP, Higashiyama R, Inoue A, et al. MicroRNA-140 is expressed in differentiated human articular chondrocytes and modulates interleukin-1 responses. Arthritis Rheum. 2009;

Chapter 4

The Function of FEN1 is Regulated by Post-Translational Modification

Zhenxing Wu, Xiaofen Mo, Chengbo Lang and Jinjing Luo

Abstract

Flap endonuclease 1 (FEN1) is a multifunctional DNA branching nuclease. Post-translational modifications (PTMs) exist in this protein widely, including phosphorylation, methylation, acetylation, ubiquitination and small ubiquitination modification (SUMO). Here, we make a summary for those PTMs studies on FEN1, to illustrate relationships between mutations of those amino acids and their functions alteration of FEN1. Numerous evidences have confirmed that dysfunction of FEN1 would lead to genome instability, and then induce a variety of chromosome-related diseases ultimately, including tumors. On one hand, interaction partner also stimulates FEN1 nuclease activity, to further ensure an effective role in the processing of different DNA structures; on the other hand, PTMs may regulate protein-protein interactions and FEN1's cellular localization.

Keywords: FEN1, post-translational modification, mutation

1. Introduction

Flap endonuclease 1 (FEN1) is one of the member of the 5′ nuclease superfamily with specific structure [1]. It could participate in Okazaki fragment maturation, removal of RNA primers in delayed strand replication, maintenance of telomeres, long base excision repair etc [2–8]. Therefore, we believe that the nuclease activity of FEN1 is to maintain genome stability, and it is a necessary condition for normal cell cycle progression. However, if FEN1 is activated on inappropriate time, it would result in interrupting the cell cycle and then cause DNA damage. Therefore, function of FEN1 must be performed at a precise location within a suitable protein complex in appropriate time. Dysregulation of FEN1 activity may lead to destruction of genetic information encoded in DNA and disrupt cell cycle. Studies have found that the expression of endogenous FEN1 increases in the G1 phase, reaches a peak in the S phase, and then drops sharply in the subsequent G2/M phase [9]. A large number of experiments have shown that FEN1 could undergo methylation [10], phosphorylation [11], SUMOylation [12, 13] and ubiquitination, a series of post-translational modifications (PTMs), finally FEN1 may degrade through the proteasome pathway in the G2/M phase [14].

2. PTMs of FEN1

2.1 Methylation and phosphorylation of FEN1

In order to complete lagging strand DNA synthesis, mammals need to efficiently and accurately process up to 50 million Okazaki fragments in each cell cycle, and researchers have found that FEN1 plays an important role [15, 16]. Henneke et al. found that methylation and phosphorylation are important signals for the binding and dissociation of FEN1 and proliferating cell nuclear antigen (PCNA) replicas [11, 17]. Methylation promotes FEN1 binding to PCNA. Interacting with PCNA, Pol-δ, replication protein A and DNA ligase I, FEN1 removes RNA primers during DNA replication [18], and then be phosphorylated, resulting in disassociation from DNA complex and entering the next step [19].

2.2 SUMO3 modification of phosphorylated FEN1

After completing the DNA synthesis of the lagging strand, phosphorylated FEN1 is dissociated from the DNA complex [19]. Hietakangas et al. and Mohideen et al. [20, 21] speculated that it was modified by SUMO3, and they used SUMO3 to measure the SUMOylation for phosphorylated FEN1 *in vitro*, and compared it with the unphosphorylated FEN1. Results showed that SUMOylation efficiency of phosphorylated FEN1 was significantly higher than that of unphosphorylated. Guo et al. co-transfected HeLa cells with a plasmid encoding His-SUMO3 and WT FEN1, S187A mutant (phosphorylated deficient) or S187D mutant (sustained phosphorylated) for FEN1's phosphorylation study [19]. Compared with WT FEN1, S187A prevented FEN1 from undergoing SUMOylation, while S187D was the opposite. Henneke et al. treated the cells with Olomoucine, SUMO3 modification of WT FEN1 and S187D FEN1 was inhibited significantly [20]. Those indicated that SUMO3 modification of FEN1 is regulated by the phosphorylation of FEN1 at S187 (**Figure 1A**).

2.3 Degradation of FEN1 mediated by SUMO3

Small ubiquitin-related modifier (SUMO) molecule is a newly discovered ubiquitin-like molecule, which participates in protein PTMs and plays an important role [22]. Mammals have SUMO1, SUMO2 and SUMO3, these 3 paralogs passed through 2 different enzymes, SUMO activating enzyme E1 and E2 [23, 24]. To test whether FEN1 degradation is related to SUMOylation or not, SUMO1, 2 and 3 overexpressing level for FEN1 was determined. It has been found that overexpression of SUMO3 led to a sharply decrease in FEN1, which indicated that SUMO3 modification is related to FEN1 degradation. SUMO3 modification site on FEN1 was identified as lysine 168 (K168). To confirm this, mutated K168R and WT FEN1 were subjected to SUMOylation *in vitro*. WT was SUMOylated, while K168R was not (**Figure 1A**). Then it was found that c-Myc-WT FEN1 was SUMOylated in cells, while SUMOylation of K168R was not. These showed that K168 is the main SUMO3 modification site of FEN1. Further research found that K168R resists its degradation of FEN1 in G2/M phase, further confirming that degradation of FEN1 requires SUMO3 modification at K168.

2.4 UBE1/UBE2M/PRP19 complex ubiquitinated FEN1

Ubiquitination is a universal signal pathway for ubiquitin-proteasome degradation for proteins [25]. After its ubiquitinated, FEN1 could be recognized as a

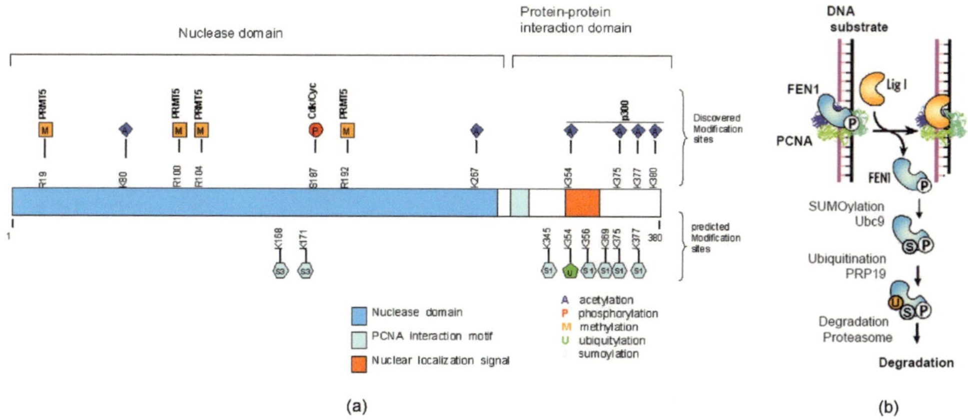

Figure 1.
(a) PTMs programmed mutation sites. Up is discovered modification sites. Down is predicted modification sites. A stands for acetylation; P stands for phosphorylation; M stands for methylation; U stands for ubiquitination; S stands for SUMOylation. (b) Model of sequential modifications to degrade FEN1. FEN1 is phosphorylated in late S phase, then dissociation from PCNA and the DNA replication fork. Once released from the DNA replication fork, phosphorylated FEN1 is then SUMOylated, which triggers ubiquitination by PRP19 and its degradation ultimately.

degradation target. Researchers determined that FEN1 is ubiquitinated *in vitro* and *in vivo*. Mass spectrometry showed that mutation of K354R abolished the ubiquitination of FEN1 (**Figure 1A**), which confirmed K354 is the ubiquitination site indeed. Ubiquitination is similar to SUMOylation, both processes involve sequential reaction steps mediated by a set of enzymes [25, 26]. Only E1 (SAE1/SAE2) and E2 (Ubc9) have been found for SUMOylation. However, it was reported that there are dozens of proteins involved in ubiquitination [27], and 138 interacting proteins then have been identified, including UBE1 [28], UBE2M [29], PRP19 [30] and UBQLN4 [31], which have been previously determined to play a role in ubiquitination pathway. It was reported that UBE1 has E1 ubiquitin-activation activity, UBE2M has E2 ubiquitin-binding activity, PRP19 has E3 ligase-activity, and UBLNQ4 may play a role in de-ubiquitination. It has also been found that the consumption of UBQLN4 had no effect on its ubiquitination, while consumption of UBE1 and UBE2M eliminated its ubiquitination, while consumption of PRP19 showed that FEN1's ubiquitination was significantly reduced. Purified UBE1, UBE2 and PRP19 were also tested, indicating that they could lead to FEN1's ubiquitination *in vitro*.

2.5 FEN1 is degraded through the ubiquitin-proteasome pathway during G2/M

It has been showed that endogenous FEN1's expression increased during G1 phase, reached a peak in S phase, and then dropped sharply in G2/M phase, which indicated that FEN1 was differentially regulated during cell cycle progression. MG132 (Carbobenzoxyl-leucinyl-leucinyl- leucinal), the proteasome inhibitor, was used for study. Results showed that the level of FEN1 in G2/M phase increased, while the level in S phase did not change significantly, suggesting that degradation is regulated by the proteasome pathway (**Figure 1B**) [32].

2.6 Acetylation of FEN1

Hasan et al. found that the transcriptional co-activator P300 histone acetyltransferase could acetylate Lys354, Lys375, Lys377 and Lys380 residues of FEN1 [33].

Acetylation of FEN1 reduced the DNA binding ability and nuclease activity significantly, and then inhibited the interaction between FEN1 and 9-1-1 complex (Rad9/Rad1/Hus1 cell cycle checkpoint complex) [34], however, binding ability to PCNA was not affected [35]. Lys80, Lys267, Lys375, have been found as 3 acetylating sites [36]. Strangely, Lys80 and Lys267 were not found in previous studies, and functions of these new sites were still unclear. New acetylation sites, such as Lysl25, Lys252, Lys254 and Lys314, have also been identified then [37]. It showed that acetylation levels of K125A and K252/K254A mutants decreased significantly, indicating that these sites were key sites for FEN1's acetylation. In addition, the proportion of G1 phase for K125A and K252/K254A increased significantly. K252/K254A mutant cells were more sensitive to UV induced DNA damage [37]. These indicated that acetylating modification of those sites plays a role in maintaining the normal physiological state of cells and their genomic stability.

2.7 Succinylation of FEN1

It has been found some succinylation modifications sites in FEN1 [38]. Lys200, Lys354 overlapped with some of identified acetylation sites, indicating that both were succinylation and acetylation modification sites for FEN1. As succinylation of lysine is more complex than acetylation, this modification may induce greater changes in chemical properties, structure and function of FEN1 [39]. Acetylation and succinylation of lysine are very important in gene transcription, cell metabolism, and DNA damage response. In addition, overlap of these modification sites for acetylation and succinylation indicated that there may be some complementary, synergistic or antagonistic effect between those modifications.

3. Summary

PTMs occur sequentially in the degradation process of FEN1, and each PTM is necessary for the next reaction. The absence or inhibition of any step would prevent FEN1 from degrading successfully. Researchers have also found that overexpression of WT FEN1 caused 25% cells closed to tetraploidy, while 65% in K354A; just only 5% in parental HeLa cells. It could be concluded that overexpression of FEN1 makes the chromosomal number in disorder. High FEN1 levels were usually observed in cancer cells. In addition, FEN1's overexpression has been observed in a variety of cancers, which is a sign for poor prognosis.

Sequential PTMs of FEN1 in the cell cycle process, is the separation of methylated FEN1 from PCNA, first undergoes phosphorylation, though demethylation reaction after methylated FEN1 has still not been illustrated. Phosphorylation induces FEN1 to undergo SUMO3 modification, and SUMO3 modification further stimulates ubiquitination. Finally, FEN1 is degraded through the proteasome pathway. Any residues of these PTMs have been mutated, degradation of FEN1 would be hindered. Therefore, there must be a precise regulation mechanism to ensure that FEN1 could function at suitable time. Numerous evidences have confirmed that the dysfunction of FEN1 would lead to instability of the genome, and induce a variety of chromosome-related diseases ultimately, including tumors. Interaction partners stimulate nuclease activity to further ensure an effective role of FEN1 in the processing of DNA structures, and PTMs also may regulate protein-protein interactions and FEN1's localization in cells. In the future, it is believed that more mechanisms of FEN1 PTMs' function would be discovered, to understand cancers development better, and it also would benefit mankind clinically.

Acknowledgements

This work was supported by the Normal Doctor Program Funding (Hangzhou Medical College, NO.2016B04). We are grateful for colleague Dr Zhigang Guo's and Dr Hong Xu's studies and their assistance, and we also appreciate for their recent reviews and studies on FEN1.

Author details

Zhenxing Wu*, Xiaofen Mo, Chengbo Lang and Jinjing Luo
School of Basic Medical Sciences and Forensic Medicine, Hangzhou Medical College, China

*Address all correspondence to: wuzhenxing007@163.com

References

[1] Satya Narayan, Aruna S Jaiswal, Brian K Law, et al. Interaction between APC and Fen1 during breast carcinogenesis. DNA Repair (Amst), 2016, 41: 54-62.

[2] U Chen, S Chen, P Saha, et al. p21Cip1/Waf1 disrupts the recruitment of human Fen1 by proliferating-cell nuclear antigen into the DNA replication complex. Proc Natl Acad Sci U S A, 1996, 93(21):11597-11602.

[3] Abhishek Saharia, Daniel C Teasley, Julien P Duxin, et al. FEN1 ensures telomere stability by facilitating replication fork re-initiation. J Biol Chem, 2010, 285(35): 27057-27066.

[4] Li Zheng, Mian Zhou, Qing Chai, et al. Novel function of the flap endonuclease 1 complex in processing stalled DNA replication forks. EMBO Rep, 2005, 6(1):83-89.

[5] Pingfang Liu, Limin Qian, Jung-Suk Sung, et al. Removal of oxidative DNA damage via FEN1-dependent long-patch base excision repair in human cell mitochondria. Mol Cell Biol, 2008, 28(16): 4975-4987.

[6] Lauren M F Merlo, John W Pepper, Brian J Reid, et al. Cancer as an evolutionary and ecological process. Nat Rev Cancer, 2006, 6(12): 924-935.

[7] Harekrushna Panda, Aruna S Jaiswal, Patrick E Corsino, et al. Amino acid Asp181 of 5'-flap endonuclease 1 is a useful target for chemotherapeutic development. Biochemistry, 2009, 48(42): 9952-9958.

[8] Lanlan Jia, Yiyi Wang, Yuejin Hua, et al. Research progress on functional regulation of flap endonuclease 1 by programmed post-translational modification. Journal of Nuclear Agricultural Sciences, 2020, 34(6): 1205-1212.

[9] Xiaoli Xu, Rongyi Shi, Li Zheng, et al. SUMO-1 modification of FEN1 facilitates its interaction with Rad9-Rad1-Hus1 to counteract DNA replication stress. J Mol Cell Biol, 2018, 10(5): 460-474.

[10] Palaniraja Thandapani, Anthony M Couturier, Zhenbao Yu, et al. Lysine methylation of FEN1 by SET7 is essential for its cellular response to replicative stress. Oncotarget, 2017, 8(39): 64918-64931.

[11] Ghislaine Henneke, Stéphane Koundrioukoff, Ulrich Hübscher. Phosphorylation of human Fen1 by cyclin-dependent kinase modulates its role in replication fork regulation. Oncogene, 2003, 22(28): 4301-4313.

[12] Xiaoli Xu, Rongyi Shi, Li Zheng, et al.SUMO-1 modification of FEN1 facilitates its interaction with Rad9-Rad1-Hus1 to counteract DNA replication stress. J Mol Cell Biol, 2018, 10(5): 460-474.

[13] Zhigang Guo, Julie Kanjanapangka, Na Liu, et al. Sequential posttranslational modifications program FEN1 degradation during cell-cycle progression. Mol Cell, 2012, 47(3): 444-456.

[14] Glickman MH, Ciechanover A. The ubiquitin-proteasome proteolytic pathway: destruction for the sake of construction. Physiol Rev, 2002, 82(2): 373-428.

[15] Burgers PM. Polymerase dynamics at the eukaryotic DNA replication fork. J Biol Chem, 2009, 284(7): 4041-4045.

[16] Garg P, Burgers PM. DNA polymerases that propagate the eukaryotic DNA replication fork. Crit Rev Biochem Mol Biol, 2005, 40(2):115-128.

[17] Yang Wu, Yingying Mei, Qinghong Gao, et al. FEN1 expression in oral

carcinoma-associated fibroblasts and its relationship with PCNA. Journal of Dental Prevention & Treatment, 2018, 26(6) : 354-359.

[18] Kang HY, Choi E, Bae SH,et al. Genetic analyses of Schizo saccharomyces pombe dna2(+) reveal that dna2 plays an essential role in Okazaki fragment metabolism. Genetics, 2000, 155: 1055-1067.

[19] Zhigang Guo, Li Zheng, Hong Xu, et al. Methylation of FEN1 suppresses nearby phosphorylation and facilitates PCNA binding. Nat Chem Biol, 2010, 6(10): 766-773.

[20] Ville Hietakangas, Julius Anckar, Henri A Blomster, et al. PDSM, a motif for phosphorylation-dependent SUMO modification. Proc Natl Acad Sci U S A, 2006, 103(1):45-50.

[21] Firaz Mohideen, Allan D Capili, Parizad M Bilimoria, et al. A molecular basis for phosphorylation-dependent SUMO conjugation by the E2 UBC9. Nat Struct Mol Biol, 2009, 16(9): 945-952.

[22] Shirong Liu, Wei He. Sumoylation-A multifunctional post-translational protein modification. Journal of Medical Molecular Biology, 2006, 3(3):212-215.

[23] Ruth Geiss-Friedlander, Frauke Melchior. Concepts in sumoylation: a decade on. Nat Rev Mol Cell Biol, 2007, 8(12):947-956.

[24] Kevin A Wilkinson, Jeremy M Henley. Mechanisms, regulation and consequences of protein SUMOylation. Review Biochem J, 2010, 428(2):133-145.

[25] Hochstrasser M. Ubiquitin-dependent protein degradation. Annu Rev Genet, 1996, 30: 405-439.

[26] Hochstrasser M. Ubiquitin, proteasomes, and the regulation of intracellular protein degradation. Review Curr Opin Cell Biol, 1995, 7(2):215-223.

[27] Pickart CM. Mechanisms underlying ubiquitination. Annu Rev Biochem, 2001, 70: 503-533.

[28] Jianping Jin, Xue Li, Steven P Gygi, et al. Dual E1 activation systems for ubiquitin differentially regulate E2 enzyme charging. Nature, 2007, 447(7148):1135-1138.

[29] Osaka F, Kawasaki H, Aida N, et al. A new NEDD8-ligating system for cullin-4A. Genes Dev, 1998, 12(15): 2263-2268.

[30] Craig W Vander Kooi, Melanie D Ohi, Joshua A Rosenberg, et al. The Prp19 U-box crystal structure suggests a common dimeric architecture for a class of oligomeric E3 ubiquitin ligases. Biochemistry, 2006, 45(1):121-130.

[31] Brigit E Riley, Yifan Xu, Huda Y Zoghbi, et al. The effects of the polyglutamine repeat protein ataxin-1 on the UbL-UBA protein A1Up. J Biol Chem, 2004, 279(40):42290-42301.

[32] Rock KL, Gramm C, Rothstein L, et al. Inhibitors of the proteasome block the degradation of most cell proteins and the generation of peptides presented on MHC class I molecules. Cell, 1994, 78(5): 761-771.

[33] S Hasan, M Stucki, P O Hassa, et al.Regulation of human flap endonuclease-1 activity by acetylation through the transcriptional coactivator p300.Mol Cell, 2001, 7(6):1221-1231.

[34] Erica Friedrich-Heineken, Magali Toueille, Barbara Tännler,et al. The two DNA clamps Rad9/Rad1/Hus1 complex and proliferating cell nuclear antigen differentially regulate flap endonuclease 1 activity. J Mol Biol, 2005, 353(5): 980-989.

[35] Erica Friedrich-Heineken, Ghislaine Henneke, Elena Ferrari, et al. The acetylatable lysines of human Fen1 are important for endo- and

exonuclease activities. J Mol Biol, 2003, 328(1): 80-84.

[36] Qijun Wang, Yakun Zhang, Chen Yang, et al. Acetylation of metabolic enzymes coordinates carbon source utilization and metabolic flux. Science, 2010, 327(5968):1004-1007.

[37] Hong Xu, Xuanyi Chen, Xiaoli Xu, et al. Lysine acetylation and succinylation in HeLa cells and their essential roles in response to UV-induced stress. Sci Rep, 2016, 26(7): 1-14.

[38] Brian T Weinert, Christian Schölz, Sebastian A Wagner,et al.Lysine succinylation is a frequently occurring modification in prokaryotes and eukaryotes and extensively overlaps with acetylation. Cell Rep, 2013, 4(4): 842-851.

[39] Matthew D Hirschey, Yingming Zhao.Metabolic regulation by lysine malonylation, succinylation and glutarylation. Mol Cell Proteomics, 2015, 14(10): 2308-2315.

Chapter 5

p300/CBP Methylation is Involved in the Potential Carcinogenic Mechanism of Lung Cancer

Yu Zhang, Wei Shen, Jin Zou and Shibo Ying

Abstract

p300/CBP is involved in the expression of a wide range of genes, both as a histone acetyltransferase (HAT) and as a coactivator of transcription factors. p300/CBP is the specific substrate of CARM1, and its KIX domain and GBD domain are the main sites methylated by arginine methyltransferase 4 (PRMT4/CARM1). p300/CBP plays an important role in lung cancer, which is a cell cycle disease. More importantly, the methylation of p300/CBP by CARM1 affects the progression of lung cancer through the cAMP-PKA pathway, p53 pathway and ER pathway. The structure, function, methylation modification sites, methylation-related enzymes, genes associated with lung cancer and the possible mechanisms of p300/CBP action are reviewed.

Keywords: p300/CBP, methylation, CARM1, signal transduction pathway, EMT, lung cancer

1. Introduction to the p300/CBP protein

1.1 Structure and function of p300/CBP

The p300/CBP molecule has at least eight functional domains. The nuclear receptor scope (RID) mainly interacts with the nuclear receptor and has a cysteine-histidine-rich domain (a CH domain), namely, the CH1 domain, also known as transcription articulation zinc finger domain 1 (TAZ1). The CH2 domain in RID includes the RING and PHD domains. The RING domain is an E3 ligase that mediates the transfer of ubiquitin to substrates by binding an E2 ubiquitin binding enzyme [1]. The PHD domain is a zinc finger domain that identifies the methylated state of histones. The CH3 domain includes ZZ ZZ-type zinc finger domain (ZZ) and transcription cohesive device zinc finger domain 2 (TAZ2). The HAT domain of p300/CBP is the structural basis of the transcription complex bridge formed by multiple transcription factors and functions as an acetylase. The structural failure of RING-HAT connections or RING domain, and particularly the loss of the RING domain, usually results in significant increases in the automatic acetylation of p300/CBP and in the acetylation of p53 [2]. The bromine domain binds to acetylated histones and transcription factors in nucleosomes. The deletion or mutation of the bromine domain does not eliminate HAT domain activity but interferes with substrate targeting and transcriptional activity [3], and bromine domain inhibitors have been shown to reduce the expression of G protein signal regulators (RGS4) [4]. The KIX domain is a CREB-binding site and the main motif modified by CARM1 methylation. Steroidal hormone receptor coactivator 1 (SID) mainly mediates protein-protein interactions, and many cell and viral proteins bind to this region, which is also the domain of the srC-1 interaction [5]. In glutamine- and proline-rich domains in the N-terminus and C-terminus have transactivation domains (TA domains), and their main function is activating transcription.

1.2 Major methylation sites of p300/CBP and related enzymes

As early as 2005, it was reported that three arginine methylation sites were found in the GBD domain (GRIP1-binding domain) in the C-terminus of p300, Arg-2056; Arg-2088; and Arg-2142. Among these residues, Arg-2142 is the most important site, and the methylation of Arg-2142 strongly inhibits the interaction between p300 and GRIP1 [6]. In subsequent studies, the importance of Arg-754 methylation in the p300·KIX domain in the cell response to DNA damage was gradually discovered, and Arg-754 in CBP is analogous to one of the three arginine residues in p300 mentioned above. The KIX domain of p300/CBP is not only the binding site of CREB but also the main site of CARM1 methylation modification, and studies have confirmed that CARM1 can be methylated to modify p300/CBP molecules both *in vivo* and *in vitro*, with Arg-754 being the main site for CARM1 methylation. It has been found that the methylation of Arg-754 can recruit the p53-binding region of BRCA1 to the p21 promoter, initiating the activation of p53 and, subsequently, p21 in response to DNA damage [7].

It has also been reported that the methylation site of CBP·KIX, Arg-580, is highly methylated by CARM1 *in vitro*, and Arg-600 of the CBP·KIX domain (equivalent to Arg-580 of p300·KIX) is located on the outer surface of the KIX-KID complex. Its methylation blocks the activation of CREB by blocking the interaction between KIX and the CREB kinase-induced domain (KID) [8]. In addition, CBP protein residues Arg714, Arg742 and Arg768 are the main methylation sites of CARM1 *in vitro*, and R742 is the main methylation site of CARM1 *in vivo* [9].

2. Expression of p300/CBP and CARM1 in lung cancer

2.1 The role of high p300/CBP expression in lung cancer tissues

Highly expressed p300 significantly enhances the ability of cancer cells to invade and migrate in non-small cell lung cancer (NSCLC). If the p300 gene is knocked out, the invasion and invasion ability of the cancer cells is significantly reduced, which may be related to the increased ZEB1 activity caused by the formation of the p300-Smad complex and further induction of the EMT [10]. CBP is highly expressed in lung cancer cells and tumor tissues. CBP acetylation is associated with cleavage and polyadenylation specificity factor subunit 4 (CPSF4) in the gene promoter region and synergically regulates downstream gene transcription and tumor cell proliferation. This association between CBP and the CPSF4 and its synergistic effect on the activation of human telomerase reverse transcriptase (hTERT) expression may contribute to the involvement of CBP in the mechanism promoting lung cancer growth [11, 12].

2.2 CARM1 is highly expressed in lung cancer tissues

Silencing CARM1 expression significantly reduced the apoptosis rate of lung cancer cells and significantly promoted the migration of lung cancer cells, while the overexpression of CARM1 significantly increased the apoptosis rate of lung cancer cells and reduced the migration of lung cancer cells, suggesting that CARM1 may attenuate the development of lung cancer [13]. Notably, a recent study found that the overexpression of CARM1 leads to methylated H3R17me2a and H3R26me2a at the core promoter region of the gene encoding the cell cycle protein E2 (CCNE2), activating CCNE2 expression at the height of the cell cycle, which facilitated the G1/S phase transition and promoted cell proliferation and colony formation. The overexpression of CCNE2 is often observed in lung cancer tissues, and the tumor-promoting effect of CARM1 in NSCLC has been shown by *in vitro* experiments [14].

Since CARM1 and PRMT1 are highly expressed in lung cancer tissues and expressed at low levels in normal tissues, CARM1 distribution was significant. Increased keratin expression and neurometalloprotein B receptor (NMBR) expression were observed in CARM1-knockout cancer cells, demonstrating that CARM1 expression is associated with lung cancer differentiation and staging [15].

3. Carcinogenic mechanism involves p300/CBP methylation

3.1 p53 pathway and $p21^{WAF1/CIP1}$ activation mechanism

p21 is a member of the KIP/CIP family of cyclin-dependent kinase (Cdk) inhibitors (CKIs) and inhibits the action of all Cdk complexes throughout the cell cycle by occupying the ATP-binding site of the Cdk complex; therefore, it is also known as cyclin inhibitory protein 1 ($p21^{WAF1/CIP1}$).

When DNA is damaged, the expression of the p53 gene is induced, and then, p53 induces $p21^{WAF1/CIP1}$ expression. The methylation of Arg-754 in the P300·KIX domain is essential for p53 activation of $p21^{WAF1/CIP1}$. First, p53 recruits CARM1-methylated p300 before it recruits BRCA1 and then mediates the interaction between the p21 promoter and the p53-binding region to initiate the activation of p53 and of p21 [7]. Activated $P21^{WAF1/CIP1}$, cyclin, Cdk and proliferating cell nuclear antigen (PCNA) combine to form the cyclin-Cdk-p21-PCNA tetramer complex, which prevents the cell proliferation signal from effectively being transmitted, and the damaged cell is arrested in the G_1 phase, inducing transcription of the DNA repair gene *GADD45*. If a damaged cell remains stagnated in the G_1 phase, p53 is induced to activate the apoptotic gene Bax and thus initiate cell apoptosis (**Figure 1a**). It has also been reported that the expression of $p21^{WAF1/CIP1}$ was directly upregulated p300/CBP-induced acetylation of KLF6 [16]. Peroxisomal proliferative factor receptor γ (PPAR-γ) also directly upregulated $p21^{WAF1/CIP1}$ expression in lung cancer cells.

3.2 cAMP-PKA pathways

As one of the most common signaling pathways of G protein-coupled receptors, the cAMP-PKA pathway plays a very important role in the regulation of cell activity. Many signaling molecules, such as glucagon, adrenaline and corticotropin, are regulated by this pathway. Under normal physiological conditions, adenylate cyclase (AC) is activated after the ligand binds specifically to the receptor, and AC converts intracellular ATP into cAMP, which is an intracellular second messenger. cAMP activates cAMP-dependent protein kinase A (PKA), and the free C-subunit of PKA encounters a specific serine residue (Ser133) site in the kinase-induced domain of cAMP reactive element-binding protein (CREB) which is phosphorylated within the nucleus to recruit p300/CBP molecules [17]. Through p300/CBP acetylation, which promotes general transcription factors (such as TFIIB) binding with the target gene promoter, target gene expression is regulated. Activated CREB has a wide range of cytological effects, including *in vitro* participation in the regulation of

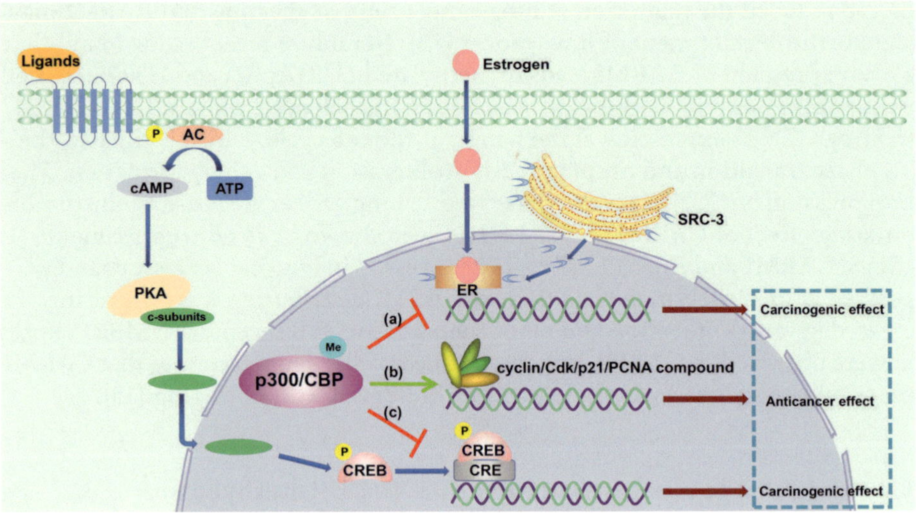

Figure 1.
a: p300/CBP methylation by CARM1 blocks the ER pathway by inhibiting the formation of core transcription complexes; b: P300/CBP methylation by CARM1 blocks cAMP-PKA-CREB activation; c: P300/CBP methylation by CARM1 is beneficial to p53-dependent p21 activation. ⊣ (red): inhibition; → (green): promotion. AC: adenylate cyclase; ATP: Adenosine triphosphate; cAMP: adenosine cyclophosphate; PKA: protein kinase A; SRC-3: steroid receptor coactivator-3; ER: estrogen receptor; Cdk: cyclin-dependent kinases; PCNA: proliferating cell nuclear antigen; CREB: cAMP-response element binding protein; CRE: cAMP response element; P: phosphorylation; Me: Methylation.

cell migration/invasion, cell proliferation, cell survival, Warburg effect induction, etc. It is involved in the immune response, tumorigenicity, vascular growth and tumor progression *in vivo* [18]. However, p300/CBP·KIX was modified by CARM1 methylation, which blocked the activation of CREB and induced apoptosis by preventing the combination of KIX and KID [8] (**Figure 1b**).

In addition, *LKB1* is the target of mutational inactivation in sporadic cancers, especially NSCLC. *LKB1* is mutated in approximately 20%-30% of NSCLC cases, making it the third most common genetic change site after *P53* and *K-RAS*. The inactivation of *LKB1* and subsequent activation of cyclic adenosine reactive element-binding protein (CREB)/CREB regulating transcription coactivator (CRTC) induced *LINC00473*. *LINC00473* is a nuclear gene that interacts with *NONO*, which is a component of the cyclic adenosine signaling pathway, to promote CRTC/CREB-mediated transcription. *LINC00473* is critical for maintaining the growth and survival of lung cancer cells [19]. Methylation of the p300/CBP·KIX domain blocks the activation of CREB and may also affect the expression of CRTC-mediated *LINC00473*, thereby blocking the progression of lung cancer.

3.3 The estrogen receptor (ER) pathway

Many steroid hormone receptors are expressed on the surface of lung cancer cells, including estrogen receptor (ER). When ligands (such as estrogen) diffuse into cells or undergo in situ synthesis and when ERs are induced to form homologous or heterologous dimers, these dimers combine with nuclear DNA enhancer ERE estrogen response elements (EHRs) and recruit steroid receptor auxiliary activation factor-3 (SRC-3). The study found that the ER compound needs two SRC-3 to form an initial stable ERE/ERα/SRC-3a/SRC-3b/p300 core complex. This is a key step in establishing the core ER coactivator complex and recruiting the p300 protein to the ER genomic binding site [20].

In subsequent studies, sequential recruitment and transcriptional activation models of the coactivators of ER were proposed: each ER monomer recruits one SRC-3, two SRC-3 molecules work together to lock one p300 molecule safely into the ER complex, and then, histone H3 acetylation is initiated. Next, CARM1 binds to the complex, where it can easily methylate its substrates SRC-3, p300, and histones. Due to the proximity of SRC-3b to ER, SRC-3a and p300, the binding of CARM1 to the ER complex results in the release of SRC-3b, and CARM1 occupies the site vacated by SRC-3b. Second, the combination with CARM1 leads to further conformational changes of p300. This structural change caused by the sequential coactivator recruitment process further alters the activity of p300 acetyltransferase and the activity of CARM1 HMT on histone H3. The synergistic effect of CARM1 and p300 enhances the acetylation of histone H3K18 and the methylation of H3R17 and promotes the synergistic activation of target gene transcription [21].

After CARM1 methylation modifies p300, the interaction between SRC and p300 is inhibited [6]. CARM1 also methylates SRC-3 and destabilizes the SRC-3/CARM1 complex [22] (**Figure 1c**). Thus, the assembly of the core ER coactivator complex (ERE/ERα/SRC-3a/SRC-3b/p300) is destroyed, and the effect of the ER pathway is blocked.

4. Discussion

As a histone acetyltransferase, p300/CBP participate in various carcinogenic signal transduction pathways through its acetylation function. Methylated p300/CBP may selectively block the transcriptional activation of cAMP-PKA and steroid-dependent pathways [23], but after CARM1 methylation modifies p300/CBP, the transmission of signaling pathways is blocked in cancer, which seems to be conducive to the suppression of the transfer of signaling and the expression of signaling pathway components, such those in the ER pathway. The methylation of p300/CBP induced by inhibiting the interaction between SRC and p300 blocks the formation of the ERE/ERα/SRC-3/p300 core complex, resulting in the inhibition of cell proliferation and cell growth. In the p53-p21 pathway, p21 recruitment by p53 is mediated by p300/CBP that has been modified by CARM1 methylation, which results in cell cycle blockade and DNA repair. In the cAMP-PKA pathway, methylated p300/CBP blocks the activation of CREB, which in turn blocks the function of CREB and inhibits cell proliferation and migration. Moreover, we speculate that the high expression of p300/CBP in lung cancer tissue may be the result of CARM1 methylation of p300/CBP, mediating the activation of cancer suppression-related signaling pathways and blocking cancer-related signaling pathways. However, the high expression of CARM1 can promote CCNE2 activation and accelerate the progression of lung cancer through the methylation of histones. Therefore, inhibiting the histone methylation of CARM1 and increasing the methylation of p300/CBP are new ideas for novel targets and the treatment of lung cancer.

5. Conclusion

CARM1 may promote the apoptosis of cancer cells and inhibit the metastasis of cancer cells through the methylation of p300/CBP. The mechanism for inhibiting the occurrence of lung cancer may involve blocking the activation of oncogenic signaling pathways and mediating the activation of tumor suppressor signaling pathways.

Acknowledgements

This study was supported by the National Natural Science Foundation of China (81973011), the Zhejiang Provincial Natural Science Foundation of China (LGF21H260010), Zhejiang Provincial Program for the Cultivation of High-level Innovative Health talents (2020), and the Science and Technology Project of Cixi City (CN2018023, CN2019031).

Conflict of interest

The authors declare no conflicts of interest.

Author details

Yu Zhang[1], Wei Shen[2], Jin Zou[1] and Shibo Ying[1*]

1 Hangzhou Medical College, Hangzhou, China

2 The Third People's Hospital of Cixi, Ningbo, China

*Address all correspondence to: shiboying@zjams.cn

References

[1] Deshaies RJ, Joazeiro CAP: RING Domain E3 Ubiquitin Ligases. Annual Review of Biochemistry 2009, 78(1):399-434.

[2] Delvecchio M, Gaucher J, Aguilar-Gurrieri C, Ortega E, Panne D: Structure of the p300 catalytic core and implications for chromatin targeting and HAT regulation. Nat Struct Mol Biol 2013, 20(9):1040-1046.

[3] Manning ET, Ikehara T, Ito T, Kadonaga JT, Kraus WL: p300 forms a stable, template-committed complex with chromatin: role for the bromodomain. Mol Cell Biol 2001, 21(12):3876-3887.

[4] Chekler EL, Pellegrino JA, Lanz TA, Denny RA, Flick AC, Coe J, Langille J, Basak A, Liu S, Stock IA et al: Transcriptional Profiling of a Selective CREB Binding Protein Bromodomain Inhibitor Highlights Therapeutic Opportunities. Chem Biol 2015, 22(12):1588-1596.

[5] Wang L, Tang Y, Cole PA, Marmorstein R: Structure and chemistry of the p300/CBP and Rtt109 histone acetyltransferases: implications for histone acetyltransferase evolution and function. Curr Opin Struct Biol 2008, 18(6):741-747.

[6] Lee YH, Coonrod SA, Kraus WL, Jelinek MA, Stallcup MR: Regulation of coactivator complex assembly and function by protein arginine methylation and demethylimination. Proc Natl Acad Sci U S A 2005, 102(10):3611-3616.

[7] Lee YH, Bedford MT, Stallcup MR: Regulated recruitment of tumor suppressor BRCA1 to the p21 gene by coactivator methylation. Genes Dev 2011, 25(2):176-188.

[8] Xu W, Chen H, Du K, Asahara H, Tini M, Emerson BM, Montminy M, Evans RM: A transcriptional switch mediated by cofactor methylation. Science 2001, 294(5551):2507-2511.

[9] Chevillard-Briet M, Trouche D, Vandel L: Control of CBP co-activating activity by arginine methylation. EMBO J 2002, 21(20):5457-5466.

[10] Hou X, Gong R, Zhan J, Zhou T, Ma Y, Zhao Y, Zhang Y, Chen G, Zhang Z, Ma S et al: p300 promotes proliferation, migration, and invasion via inducing epithelial-mesenchymal transition in non-small cell lung cancer cells. BMC Cancer 2018, 18(1):641.

[11] Tang D, Zhao YC, Qian D, Liu H, Luo S, Patz EF, Moorman PG, Su L, Shen S, Christiani DC et al: Novel genetic variants in HDAC2 and PPARGC1A of the CREB-binding protein pathway predict survival of non-small-cell lung cancer. Mol Carcinog 2020, 59(1):104-115.

[12] Tang Z, Yu W, Zhang C, Zhao S, Yu Z, Xiao X, Tang R, Xuan Y, Yang W, Hao J et al: CREB-binding protein regulates lung cancer growth by targeting MAPK and CPSF4 signaling pathway. Mol Oncol 2016, 10(2):317-329.

[13] Hu B, Li X, Chen L, Liu Z: High Expression of CARM1 Inhibits Lung Cancer Progression by Targeting TP53 by Regulating CTNNB1. Lung 2020, 198(2):415-422.

[14] Wu D, He J, Zhang W, Wang K, Jin S, Li J, Gao W: CARM1 promotes non-small cell lung cancer progression through upregulating CCNE2 expression. Aging (Albany NY) 2020, 12(11):10578-10593.

[15] Elakoum R, Gauchotte G, Oussalah A, Wissler MP, Clement-Duchene C, Vignaud JM, Gueant JL, Namour F: CARM1 and PRMT1 are dysregulated in lung cancer

without hierarchical features. Biochimie 2014, 97:210-218.

[16] Li D, Yea S, Dolios G, Martignetti JA, Narla G, Wang R, Walsh MJ, Friedman SL: Regulation of Kruppel-like factor 6 tumor suppressor activity by acetylation. Cancer Res 2005, 65(20):9216-9225.

[17] Clark MD, Kumar GS, Marcum R, Luo Q, Zhang Y, Radhakrishnan I: Molecular Basis for the Mechanism of Constitutive CBP/p300 Coactivator Recruitment by CRTC1-MAML2 and Its Implications in cAMP Signaling. Biochemistry 2015, 54(35):5439-5446.

[18] Steven A, Seliger B: Control of CREB expression in tumors: from molecular mechanisms and signal transduction pathways to therapeutic target. Oncotarget 2016, 7(23):35454-35465.

[19] Chen Z, Li JL, Lin S, Cao C, Gimbrone NT, Yang R, Fu DA, Carper MB, Haura EB, Schabath MB et al: cAMP/CREB-regulated LINC00473 marks LKB1-inactivated lung cancer and mediates tumor growth. J Clin Invest 2016, 126(6):2267-2279.

[20] Yi P, Wang Z, Feng Q, Pintilie GD, Foulds CE, Lanz RB, Ludtke SJ, Schmid MF, Chiu W, O'Malley BW: Structure of a biologically active estrogen receptor-coactivator complex on DNA. Mol Cell 2015, 57(6):1047-1058.

[21] Yi P, Wang Z, Feng Q, Chou CK, Pintilie GD, Shen H, Foulds CE, Fan G, Serysheva I, Ludtke SJ et al: Structural and Functional Impacts of ER Coactivator Sequential Recruitment. Mol Cell 2017, 67(5):733-743 e734.

[22] Feng Q, Yi P, Wong J, O'Malley BW: Signaling within a coactivator complex: methylation of SRC-3/AIB1 is a molecular switch for complex disassembly. Mol Cell Biol 2006, 26(21):7846-7857.

[23] Naeem H, Cheng D, Zhao Q, Underhill C, Tini M, Bedford MT, Torchia J: The activity and stability of the transcriptional coactivator p/CIP/SRC-3 are regulated by CARM1-dependent methylation. Mol Cell Biol 2007, 27(1):120-134.

Chapter 6

The Role of Protein Arginine Methyltransferase 1 in Gastrointestinal Cancers

Jin Zou, Wei Shen, Yu Zhang and Shibo Ying

Abstract

Mammals can produce nine kinds of arginine methylation enzymes that can be divided into three types (I, II, and III) according to their catalytic activity. Arginine methyltransferase 1 (PRMT1), as the first discovered arginine methyltransferase type I, has been reported to be involved in cell signal transduction, DNA damage repair, RNA transcription and other processes. Its imbalance or abnormal expression is also involved in cancer metastasis. PRMT1 is highly expressed in gastrointestinal tumors and promotes tumor biomarkers expression, chemotherapy resistance and tumorigenicity to promote cancer progression, while downregulation of PRMT1 expression can inhibit the migration and invasion of related tumor cells or promote tumor cells apoptosis and inhibit the progression of cancer. Therefore, PRMT1 may be a cancer therapeutic target. In this paper, arginine methylase 1 expression in various types of gastrointestinal tumors, the tumorigenic mechanism and the role of PRMT1 in tumorigenesis and development were reviewed.

Keywords: PRMT1, gastrointestinal cancers, arginine methylation

1. Introduction

1.1 Protein arginine methylation

Arginine methylation is a common type of protein posttranslational modification (PTM) that preserves arginine's positive charge but reduces its hydrogen bonding capacity because each methyl group removes a hydrogen atom. Moreover, methylation increases the hydrophobicity of the side chain, thus facilitating the interaction with the aromatic ring [1]. Arginine methylation leads to changes in gene expression by altering the nucleoprotein-DNA interaction. Arginine can occur monomethylation (MMA), symmetric dimethylation (SDMA) and asymmetric dimethylation (ADMA) under the catalysis of different protein arginine methylases (PRMTs). At present, the methyl arginine identified in eukaryotes mainly occurs in three types: ω-N^G-methylarginine (MMA), ω-N^G,N^G-asymmetric dimethyl arginine (aDMA) and ω-N^G,N'^G-symmetric dimethyl arginine (sDMA) [2, 3]. Protein arginine methylation affects many important biological pathways and plays a key role in DNA damage signal transduction, pre-mRNA splicing, mRNA translation, cell signal transduction and cell fate determination [3].

IntechOpen

1.2 Protein arginine methyltransferase

PRMT catalyzes the transfer of methyl from s-adenosine methionine (SAM) to arginine arc nitrogen atoms to produce s-adenosine homocysteine and methylarginine in histones and non-histones [3]. PRMT of histone methylation and histones, according to their different catalytic activities, can be divided into type I (PRMT1, PRMT2, PRMT3, PRMT4/CARM1, PRMT6 and PRMT8), type II (PRMT5 and PRMT9) and type III (PRMT7). PRMT is a highly conserved gene product that plays a major role in normal body development and disease. In most cases, the expression of PRMT was upregulated. Maladjustment or abnormal expression of PRMT influences the development of cancer, especially the overexpression of PRMT1, PRMT4 and PRMT5, which has been confirmed in many malignant tumors [4, 5]. PRMT is associated with a variety of diseases, such as tumors, cardiovascular diseases, viral infections, and autoimmune diseases [6]. Studies have shown that PRMT can be a potentially interesting therapeutic target [5].

1.3 Protein arginine methyltransferase 1

As arginine methylation is closely related to various tumors, more and more researchers are beginning to study the relationship between PRMT and cancer, especially PRMT1. PRMT1, PRMT3, PRMT6 and PRMT8 were all highly expressed in arginine methyltransferase, but the expression of PRMT1 was significantly upregulated [7]. PRMT1 was the first arginine methylase discovered and is the major type I enzyme in mammals [8, 9]; furthermore, PRMT1 is responsible for monomethylation and more than 80% of ADMA modifications [5]. As PRMT1 activity was lost, the MMA and SDMA levels increased significantly [10]. The expression of PRMT1 in cancer cells of various tissues was significantly higher than that in nonneoplastic cells [11], and the expression level in embryonic nerve tissues was the highest [12]. PRMT1 has been found to be overexpressed or abnormally spliced in malignant tumors such as those of the breast, prostate, lung, colon, bladder and leukemia. Previous studies have also found that PRMT1 is an Important adjustment factor of epithelial-mesenchymal transition (EMT) [13, 14]. In contrast to the PRMT5 symmetrical methylated histone H4R3me2s involved in transcriptional inhibition, the PRMT1 asymmetrically methylated histone H4R3me2a recruits the p300/cAMP-binding-protein (p300/CBP) related factor complex, enhances histone H3 acetylation in lysines 9 and 14, promotes transcription factors binding, and participates in transcriptional activation [3, 7]. H4R3 methylation causes p300 to acetylate the H4 tail, while PRMT1 inhibits acetylation of the H4 tail [15]. Studies have shown that only PRMT1 and EGFR 2 (D2) coincubated with colon cancer methylation screening tests produced strong methylation signals *in vitro* [16]. PRMT3 overexpression does not regulate HBV transcription, while PRMT1 overexpression leads to HBV transcriptional inhibition [17]. This paper mainly describes the role of PRMT1 in gastrointestinal tumorss.

2. PRMT1 in cancers

2.1 PRMT1 in esophageal cancer

In 2015, Virendra Singh et al. reported for the first time that PRMT1 was involved in the transition from low to high degree of tumor formation in esophageal cancer (EC) When ESCC was poorly differentiated, moderately differentiated and then well differentiated, the expression of PRMT1 decreased [7]. PRMT1 was found in 89.5% of ESCC patients and in only 46.3% of adjacent normal tissues, and the expression level of PRMT1 in ESCC cell lines was significantly upregulated compared with that in normal esophageal epithelial cell lines [18]. The overexpression of PRMT1 led to the proliferation of OV6 + ECA109 and TE1 cells, while the downregulation of PRMT1 reduced the tumorigenicity and tumor growth of OV6+ cells. Xenotransplantation of NOD/SCID mice showed that PRMT1 expression enhanced the tumorigenicity of OV6 + ESCC cells *in vivo* [18]. Further studies showed that PRMT1 inhibited H3K9 methylation by catalyzing H4R3me2a methylation and promoted acetylation of H3 lysine residues, which enhanced chromatin activity and resulted in increased ESCC transcription [7]. PRMT1 upregulated histone H4R3me2a expression, promoted TIC markers, stem cell-like properties, chemotherapy resistance, and oncogenic expression, and increased PRMT1 expression in ECSS samples. In addition, RNA-seq transcriptome analysis showed that PRMT1 overexpression led to activation of the Wnt/β-catenin and Notch signaling pathways [18]. In conclusion, as a new effector, the PRMT1 expression level is closely related to abnormal clinicopathological features and poor patient prognosis, and PRMT1 may be a reliable diagnostic and therapeutic target for esophageal cancer.

2.2 PRMT1 in gastric carcinoma

Currently, PRMT1 and FOXO1 are mainly expressed in the nucleus of gastric cancer (GC) cell lines, and FOXO1 expression is correlated with the PRMT1 level. PRMT1 may regulate chemotherapy sensitivity and apoptosis of GC cells by activating the tumor suppressors FOXO1 and BAD [19]. Interestingly, PRMT1 inhibited drug resistance and nuclear accumulation of p-FOXO1 and p-BAD in GC cell lines, and the recurrence rate of GC in patients with low expression of PRMT1 after adjuvant chemotherapy was significantly higher than GC in patients with high expression of PRMT1. Cisplatin and 5-flu-orouracil sensitivities were inhibited by RNA interference with PRMT1 downregulation in GC cells [19]. After that, other studies suggested that PRMT1 overexpression in GC cells had the effect of "migration-proliferation", which could promote the migration and invasion and inhibit the proliferation of tumor cells, while PRMT1 knockdown had the opposite effect [13]. PRMT1 is a novel regulator of EMT that is reported to enhance migration and invasion by Hippo signaling and promote EMT. PRMT1 can reduce the expression of E-cadherin, the epithelial marker of GC, and increase the expression of the interstitial markers N-cadherin, Vimentin, Snail and Catenin [13]. In conclusion, evaluating the expression of PRMT1 in GC is an effective predictor of poor prognosis and recurrence after adjuvant chemotherapy. However, in view of its dual functions, caution should be taken prior to utilizing PRMT1 as a potential drug target for GC.

2.3 PRMT1 in colorectal cancer

Colorectal cancer (CRC) is a common malignant tumor in the gastrointestinal tract. PRMT1 is overexpressed in colorectal adenoma, carcinoma and adenocarcinoma, and the expression level of PRMT1 in colon cancer samples is higher than normal colon and rectal samples [20]. Compared with normal tissue, the expression of the PRMT1-v1 variant was significantly increased in colon cancer tissue and increased as normal tissue progressed to adenoma and eventually to cancer. In other words, the higher the degree of malignancy, the higher the expression of the variant. The Cox proportional hazard regression model and Kaplan–Meier method showed that patients with high expression of PRMT1-v1 variants had a higher probability of recurrence or death and a lower survival probability [21]. After PRMT1 was knocked out, the proliferation of HCT116 cells was significantly inhibited, and the apoptosis rate was increased. Treating HCT116 cells with downregulated PRMT1 with sodium propionate inhibited the mTOR signaling pathway to induce cell apoptosis, thereby inhibiting cell growth and proliferation [20]. PRMT1 methylates epidermal growth factor receptor (EGFR) in the extracellular region of the endoplasmic reticulum/Golgi body, enhancing ligand binding and receptor activation before transport to the cell membrane. PRMT1 mainly methylates the EGFR extracellular domain at R198 and R200; enhances the binding to EGF and the subsequent receptor dimerization and signal transduction activation; enhances the receptor function of CRC cells; promotes the growth of EGFR-dependent cells; and reduces cell resistance to cetuximab. When PRMT1 is knocked out, the EGFR methylation signal is reduced [16]. In conclusion, PRMT1 can be considered a useful therapeutic marker for the treatment of CRC, and the development of new methods to downregulate the expression of PRMT1 is of great significance for the prognosis and treatment of CRC.

2.4 PRMT1 in hepatocellular carcinoma

Increasing evidence shows that PRMT1 expression in clinical hepatocellular carcinoma (HCC) samples and cell lines is significantly higher than adjacent normal liver tissue, and high PRMT1 expression is closely related to poor prognosis and recurrence of HCC. PRMT1 upregulation in HCC cell lines promoted cell proliferation, colony formation and migration *in vitro*, while the knockdown of the PRMT1 gene inhibited that role [14, 22]. Bingshou Li et al. found that the high expression of PRMT1 was associated with the low expression of miR-503. MiR-503 can inhibit the invasion and migration of HCC cells by targeting the 3'-UTR of the PRMT1 gene, resulting in downregulation of the mRNA and protein expression of PRMT1 [23]. Further studies found that PRMT1 knockdown resulted in increased hepatocyte proliferation and decreased Hnf4a expression. In the absence of PRMT1, JMJD6 causes the Hnf4a promoter to undergo arginine demethylation, leading to the significant downregulation of Hnf4a expression and the promotion of hepatocyte proliferation. Knockout of JMJD6 restored Hnf4a expression and inhibited hepatocyte proliferation in PRMT1-knockout mice [24]. In addition, PRMT1 can also increase STAT3 phosphorylation through high expression and activate the STAT3 signaling pathway to promote *in vitro* and *in vivo* metastasis of HCC cells, while cryptotanshinone, a STAT3 inhibitor, inhibits STAT3 phosphorylation and inhibits HCC proliferation and migration [22]. Similar to PRMT1 in GC, PRMT1 is also associated with EMT in liver cancer. The expression of PRMT1 downregulated TGF-β1, p-Smad2 and p-Smad3; significantly reduced expression of the interstitial markers Vimentin, Snail and N-cadherin; and upregulated the expression of the epithelial marker E-cadherin. PRMT1 overexpression leads to the opposite effect. Therefore, PRMT1 may promote EMT in HCC cells through the TGF-β1/Smad pathway [14]. PRMT1 is also a negative adjustment factor of HBV transcription. Studies have shown that overexpression of PRMT1 in HepG2 cells results in inhibition of 60% HBV transcription, and low expression of PRMT1 significantly increased HBV transcription by 1.6-fold. *In vivo* animal models, PRMT1 activity was further reduced in HBV-replicating cells. HBx binding to PRMT1 may facilitate HBV replication [17]. In conclusion, PRMT1 may be a new therapeutic target for liver cancer prognosis, which is of great significance for improving therapeutic strategies for HCC patients.

2.5 PRMT1 in pancreatic cancer

In 2018, Zhibin Lin et al. found that PRMT1 was abnormally upregulated in permanent pancreatic cancer (PC) cell lines and human pancreatic tumors compared with nonneoplastic pancreatic epithelial tissues, but the effect of PRMT1 upregulation on tumor cells is controversial. In PANC-1 and SW1990 cells, PRMT1 overexpression did not affect the proliferation or invasion of PC cells [25]. Interestingly, the latest research data show that PRMT1 has a protumor effect and that inhibiting PRMT1 inhibits tumor growth *in vivo* and *in vitro*. PRMT1 expression is upregulated in human pancreatic ductal adenocarcinoma (PDAC) and promotes the growth and carcinogenesis of PC cells both *in vivo* and *in vitro* and is associated with higher tumor grade, higher invasivity, and poorer prognosis [26, 27]. PRMT1 downregulation significantly inhibited tumor proliferation and invasion *in vitro* and *in vivo*. Further studies found that ZEB1 expression in PC cells was closely related to PRMT1 expression, that ZEB1 expression was inhibited in PC cells downregulated by PRMT1, and that overexpression of ZEB1 reversed the antitumor effect of PRMT1 downregulation [25]. PRMT1 may act as a positive regulator of β-catenin, increasing the cellular β-catenin levels. The overexpression of exogenous PRMT1 in PC cells promoted the growth of tumor cells and increased the β-catenin levels after treatment with lithium chloride. PRMT1 expression downregulation inhibited the growth and tumorigenicity of pancreatic cancer cells and inhibited the increase in β-catenin protein under lithium chloride treatment [27]. PRMT1 overexpression enhances HSP70 binding and BCL-2 mRNA stability through the gold-rich element in the 3'-UTR, thereby increasing BCL-2 protein expression and protecting cancer cells from cell stress and treatment-induced apoptosis. The PRMT1 inhibitors DB75 and TC-E5003 reduce PRMT1-mediated protein arginine methylation, thereby inhibiting PRMT1 enzyme activity but not its expression level [26]. The levels of total Gli1 and methylated Gli1 were positively correlated with PRMT1 protein levels in human PDAC specimens. PRMT1 methylates the oncogenic transcription factor Gli1 in R597 to enhance transcriptional activity by enhancing the binding of Gli1 to its target gene promoter, while disruption of Gli1 methylation weakens the oncogenic function of Gli1 and sensitizes PDAC cells to gemcitabine therapy [28]. Downregulation PRMT1 was associated with the PD-L1 downregulation. The inhibitor PT1001B enhanced the inhibition of anti-PD-L1 on tumor cell proliferation and enhanced the induction of tumor cell apoptosis. Therefore, the combination of a protein arginine methyltransferase inhibitor (PD-1) and anti-programmed death ligand-1 (PD-L1) can effectively inhibit the progression of PC [29]. In conclusion, PRMT1 may serve as a potential biomarker for pancreatic cancer.

3. Conclusion

Protein arginine methylation affects many important biological pathways, such as DNA damage signal transduction, pre-mRNA splicing, mRNA translation, and cell signal transduction [3]. More and more evidences have shown that arginine methyltransferase is involved in various physiological and pathological processes in humans, especially in malignant tumors. Studies have found that PRMT1 is involved in the development and diseases of the nervous system and plays an important role in neurodegenerative diseases [30]. PRMT1 also promotes asthma by regulating asthma-related pri-let-7i and pri-miR-423 [31]. PRMT1-v2 activated the gluconeogenic program in hepatocytes via interactions with PGC1α, a key transcriptional coactivator regulating gluconeogenesis [32]. PRMT1 is involved in the progression of lung cancer by regulating the high expression of FEN1 [33]. PRMT1 can also promote the metastasis of breast cancer by regulating the expression of EZH2 [34]. In glioma cells, upregulation of PRMT1 can promote the growth and metastasis of glioma cells, and downregulation of PRMT1 can also produce opposite inhibition [35]. PRMT1, as the main type I enzyme in mammals [8, 9], is responsible for arginine mono methylation and more than 80% asymmetric methylation modifications [5].

In gastrointestinal tumors, PRMT1 expression has been proved to be upregulated, and its imbalance or abnormal expression is involved in the occurrence and development of cancer. Current research evidence shows that PRMT1 plays a tumorigenic role in gastrointestinal tumors. PRMT1 upregulation can promote the growth and proliferation of EC cells [1, 2], CRC cells [6, 7, 9], and PC cells [15–19]. In GC cells, PRMT1 upregulation promotes tumor cell migration, invasion and mesenchymal transformation of epithelial cells and inhibits GC cell proliferation [3, 5]. In HCC cells, PRMT1 is upregulated to promote tumor cell proliferation, migration, invasion and mesenchymal transformation of epithelial cells [10–14]. However, PRMT1 knockdown or deletion may have the opposite effect. Therefore, PRMT1 may be used as a new potential tumor biomarker and target for prognosis therapy. At present, many PRMT1 inhibitors, such as AMI-1, MS023 and GSK3368715, have entered the first phase of clinical trials, trying to open up a new way of cancer treatment [1, 36]. The latest animal experimental results show that the growth rate of HT-29 tumor cell line after xenotransplantation is slowed down under the treatment of the inhibitor MS023 [37]. Of course, there are some new inhibitors of PRMT1 under study. The latest research results show that the inhibitor TC-E-5003 has a good inhibitory effect on lung cancer and breast cancer and can also be used as an antitumor drug [38]. However, the role of PRMT1 in prognostic therapy needs to be further studied.

Acknowledgements

This study was supported by the National Natural Science Foundation of China (81973011), Basic and Public Welfare Research Project of Zhejiang Province (LGF21H260010), Zhejiang Provincial Program for the Cultivation of High-level Innovative Health talents (2020), and the Science and Technology Project of Cixi City (CN2018023, CN2019031).

Conflict of interest

The authors declare no conflicts of interest.

Author details

Jin Zou[1], Wei Shen[2], Yu Zhang[1] and Shibo Ying[1]*

1 Hangzhou Medical College, Hangzhou, China

2 The Third People's Hospital of Cixi, Ningbo, China

*Address all correspondence to: shiboying@zjams.cn

References

[1] Guccione E, Richard S. The regulation, functions and clinical relevance of arginine methylation. Nat Rev Mol Cell Biol. 2019 Oct;20(10):642-657. doi: 10.1038/s41580-019-0155-x. Epub 2019 Jul 26. Erratum in: Nat Rev Mol Cell Biol. 2019 Sep;20(9):567. PMID: 31350521.

[2] Paik WK, Paik DC, Kim S. Historical review: the field of protein methylation. Trends Biochem Sci. 2007 Mar;32(3):146-52. doi: 10.1016/j.tibs.2007.01.006. Epub 2007 Feb 8. PMID: 17291768.

[3] Blanc RS, Richard S. Arginine Methylation: The Coming of Age. Mol Cell. 2017 Jan 5;65(1):8-24. doi: 10.1016/j.molcel.2016.11.003. PMID: 28061334.

[4] Yang Y, Bedford MT. Protein arginine methyltransferases and cancer. Nat Rev Cancer. 2013 Jan;13(1):37-50. doi: 10.1038/nrc3409. Epub 2012 Dec 13. PMID: 23235912.

[5] Smith E, Zhou W, Shindiapina P, Sif S, Li C, Baiocchi RA. Recent advances in targeting protein arginine methyltransferase enzymes in cancer therapy. Expert Opin Ther Targets. 2018 Jun; 22(6):527-545. doi: 10.1080/14728222.2018.1474203. Epub 2018 May 21. PMID: 29781349; PMCID: PMC6311705.

[6] Peng C, Wong CC. The story of protein arginine methylation: characterization, regulation, and function. Expert Rev Proteomics. 2017 Feb;14(2):157-170. doi: 10.1080/14789450.2017.1275573. Epub 2017 Jan 5. PMID: 28043171.

[7] Singh V, Singh LC, Singh AP, Sharma J, Borthakur BB, Debnath A, Rai AK, Phukan RK, Mahanta J, Kataki AC, Kapur S, Saxena S. Status of epigenetic chromatin modification enzymes and esophageal squamous cell carcinoma risk in northeast Indian population. Am J Cancer Res. 2015 Feb 15;5(3):979-99. PMID: 26045981; PMCID: PMC4449430.

[8] Zhu L, He X, Dong H, Sun J, Wang H, Zhu Y, Huang F, Zou J, Chen Z, Zhao X, Li L. Protein arginine methyltransferase 1 is required for maintenance of normal adult hematopoiesis. Int J Biol Sci. 2019 Oct 23;15(13):2763-2773. doi: 10.7150/ijbs.38859. PMID: 31853216; PMCID: PMC6909962.

[9] Tang J, Frankel A, Cook RJ, Kim S, Paik WK, Williams KR, Clarke S, Herschman HR. PRMT1 is the predominant type I protein arginine methyltransferase in mammalian cells. J Biol Chem. 2000 Mar 17;275(11):7723-30. doi: 10.1074/jbc.275.11.7723. PMID: 10713084.

[10] Dhar S, Vemulapalli V, Patananan AN, Huang GL, Di Lorenzo A, Richard S, Comb MJ, Guo A, Clarke SG, Bedford MT. Loss of the major Type I arginine methyltransferase PRMT1 causes substrate scavenging by other PRMTs. Sci Rep. 2013;3:1311. doi: 10.1038/srep01311. PMID: 23419748; PMCID: PMC3575585.

[11] Yoshimatsu M, Toyokawa G, Hayami S, Unoki M, Tsunoda T, Field HI, Kelly JD, Neal DE, Maehara Y, Ponder BA, Nakamura Y, Hamamoto R. Dysregulation of PRMT1 and PRMT6, Type I arginine methyltransferases, is involved in various types of human cancers. Int J Cancer. 2011 Feb 1;128(3):562-73. doi: 10.1002/ijc.25366. PMID: 20473859.

[12] Pawlak MR, Scherer CA, Chen J, Roshon MJ, Ruley HE. Arginine N-methyltransferase 1 is required for early postimplantation mouse development, but cells deficient in the enzyme are viable. Mol Cell Biol. 2000 Jul;20(13):4859-69. doi: 10.1128/mcb.20.13.4859-4869.2000. PMID: 10848611; PMCID: PMC85937.

[13] Zhang Y, Wang D, Zhang M, Wei H, Lu Y, Sun Y, Zhou M, Gu S, Feng W,

Wang H, Zeng J, Gong A, Xu M. Protein arginine methyltransferase 1 coordinates the epithelial-mesenchymal transition/proliferation dichotomy in gastric cancer cells. Exp Cell Res. 2018 Jan 1;362(1):43-50. doi: 10.1016/j.yexcr.2017.10.035. Epub 2017 Oct 31. PMID: 29097184.

[14] Wei H, Liu Y, Min J, Zhang Y, Wang J, Zhou M, Xiong E, Yu G, Zhou H, He J, Zeng J, Gong A, Xu M. Protein arginine methyltransferase 1 promotes epithelial-mesenchymal transition via TGF-β1/Smad pathway in hepatic carcinoma cells. Neoplasma. 2019 Nov;66(6):918-929. doi: 10.4149/neo_2018_181226N999. Epub 2019 Oct 8. PMID: 31607134.

[15] Wang H, Huang ZQ, Xia L, Feng Q, Erdjument-Bromage H, Strahl BD, Briggs SD, Allis CD, Wong J, Tempst P, Zhang Y. Methylation of histone H4 at arginine 3 facilitating transcriptional activation by nuclear hormone receptor. Science. 2001 Aug 3;293(5531):853-7. doi: 10.1126/science.1060781. Epub 2001 May 31. PMID: 11387442.

[16] Liao HW, Hsu JM, Xia W, Wang HL, Wang YN, Chang WC, Arold ST, Chou CK, Tsou PH, Yamaguchi H, Fang YF, Lee HJ, Lee HH, Tai SK, Yang MH, Morelli MP, Sen M, Ladbury JE, Chen CH, Grandis JR, Kopetz S, Hung MC. PRMT1-mediated methylation of the EGF receptor regulates signaling and cetuximab response. J Clin Invest. 2015 Dec;125(12):4529-43. doi: 10.1172/JCI82826. Epub 2015 Nov 16. PMID: 26571401; PMCID: PMC4665782.

[17] Benhenda S, Ducroux A, Rivière L, Sobhian B, Ward MD, Dion S, Hantz O, Protzer U, Michel ML, Benkirane M, Semmes OJ, Buendia MA, Neuveut C. Methyltransferase PRMT1 is a binding partner of HBx and a negative regulator of hepatitis B virus transcription. J Virol. 2013 Apr;87(8):4360-71. doi: 10.1128/JVI.02574-12. Epub 2013 Feb 6. PMID: 23388725; PMCID: PMC3624337.

[18] Zhao Y, Lu Q, Li C, Wang X, Jiang L, Huang L, Wang C, Chen H. PRMT1 regulates the tumour-initiating properties of esophageal squamous cell carcinoma through histone H4 arginine methylation coupled with transcriptional activation. Cell Death Dis. 2019 May 1;10(5):359. doi: 10.1038/s41419-019-1595-0. PMID: 31043582; PMCID: PMC6494844.

[19] Altan B, Yokobori T, Ide M, Mochiki E, Toyomasu Y, Kogure N, Kimura A, Hara K, Bai T, Bao P, Suzuki M, Ogata K, Asao T, Nishiyama M, Oyama T, Kuwano H. Nuclear PRMT1 expression is associated with poor prognosis and chemosensitivity in gastric cancer patients. Gastric Cancer. 2016 Jul;19(3):789-97. doi: 10.1007/s10120-015-0551-7. Epub 2015 Oct 15. PMID: 26472729.

[20] Ryu TY, Kim K, Son MY, Min JK, Kim J, Han TS, Kim DS, Cho HS. Downregulation of PRMT1, a histone arginine methyltransferase, by sodium propionate induces cell apoptosis in colon cancer. Oncol Rep. 2019 Mar;41(3):1691-1699. doi: 10.3892/or.2018.6938. Epub 2018 Dec 18. PMID: 30569144; PMCID: PMC6365698.

[21] Mathioudaki K, Papadokostopoulou A, Scorilas A, Xynopoulos D, Agnanti N, Talieri M. The PRMT1 gene expression pattern in colon cancer. Br J Cancer. 2008 Dec 16;99(12):2094-9. doi: 10.1038/sj.bjc.6604807. PMID: 19078953; PMCID: PMC2607217.

[22] Zhang XP, Jiang YB, Zhong CQ, Ma N, Zhang EB, Zhang F, Li JJ, Deng YZ, Wang K, Xie D, Cheng SQ. PRMT1 Promoted HCC Growth and Metastasis *In Vitro* and *In Vivo* via Activating the STAT3 Signalling Pathway. Cell Physiol Biochem. 2018;47(4):1643-1654. doi: 10.1159/000490983. Epub 2018 Jun 23. PMID: 29945155.

[23] Li B, Liu L, Li X, Wu L. miR-503 suppresses metastasis of hepatocellular carcinoma cell by targeting PRMT1. Biochem Biophys Res Commun. 2015 Sep 4;464(4):982-987. doi: 10.1016/j.bbrc.2015.06.169. Epub 2015 Jul 7. PMID: 26163260.

[24] Zhao J, Adams A, Roberts B, O'Neil M, Vittal A, Schmitt T, Kumer S, Cox J, Li Z, Weinman SA, Tikhanovich I. Protein arginine methyl transferase 1- and Jumonji C domain-containing protein 6-dependent arginine methylation regulate hepatocyte nuclear factor 4 alpha expression and hepatocyte proliferation in mice. Hepatology. 2018 Mar;67(3):1109-1126. doi: 10.1002/hep.29587. Epub 2018 Jan 24. PMID: 29023917; PMCID: PMC5826837.

[25] Lin Z, Chen Y, Lin Z, Chen C, Dong Y. Overexpressing PRMT1 Inhibits Proliferation and Invasion in Pancreatic Cancer by Inverse Correlation of ZEB1. IUBMB Life. 2018 Oct;70(10):1032-1039. doi: 10.1002/iub.1917. Epub 2018 Sep 8. PMID: 30194893.

[26] Wang L, Jia Z, Xie D, Zhao T, Tan Z, Zhang S, Kong F, Wei D, Xie K. Methylation of HSP70 Orchestrates Its Binding to and Stabilization of BCL2 mRNA and Renders Pancreatic Cancer Cells Resistant to Therapeutics. Cancer Res. 2020 Oct 15;80(20):4500-4513. doi: 10.1158/0008-5472.CAN-19-1738. Epub 2020 Jul 22. PMID: 32699135.

[27] Song C, Chen T, He L, Ma N, Li JA, Rong YF, Fang Y, Liu M, Xie D, Lou W. PRMT1 promotes pancreatic cancer growth and predicts poor prognosis. Cell Oncol (Dordr). 2020 Feb;43(1):51-62. doi: 10.1007/s13402-019-00435-1. Epub 2019 Sep 13. Erratum in: Cell Oncol (Dordr). 2019 Dec 2;: PMID: 31520395.

[28] Wang Y, Hsu JM, Kang Y, Wei Y, Lee PC, Chang SJ, Hsu YH, Hsu JL, Wang HL, Chang WC, Li CW, Liao HW, Chang SS, Xia W, Ko HW, Chou CK, Fleming JB, Wang H, Hwang RF, Chen Y, Qin J, Hung MC. Oncogenic Functions of Gli1 in Pancreatic Adenocarcinoma Are Supported by Its PRMT1-Mediated Methylation. Cancer Res. 2016 Dec 1;76(23):7049-7058. doi: 10.1158/0008-5472.CAN-16-0715. Epub 2016 Oct 6. PMID: 27758883; PMCID: PMC5135656.

[29] Zheng NN, Zhou M, Sun F, Huai MX, Zhang Y, Qu CY, Shen F, Xu LM. Combining protein arginine methyltransferase inhibitor and anti-programmed death-ligand-1 inhibits pancreatic cancer progression. World J Gastroenterol. 2020 Jul 14;26(26):3737-3749. doi: 10.3748/wjg.v26.i26.3737. PMID: 32774054; PMCID: PMC7383845.

[30] Hashimoto M, Fukamizu A, Nakagawa T, Kizuka Y. Roles of protein arginine methyltransferase 1 (PRMT1) in brain development and disease. Biochim Biophys Acta Gen Subj. 2021 Jan;1865(1):129776. doi: 10.1016/j.bbagen.2020.129776. Epub 2020 Oct 28. PMID: 33127433.

[31] Zhai W, Sun H, Li Z, Li L, Jin A, Li Y, Chen J, Yang X, Sun Q, Lu S, Roth M. PRMT1 Modulates Processing of Asthma-Related Primary MicroRNAs (Pri-miRNAs) into Mature miRNAs in Lung Epithelial Cells. J Immunol. 2020 Nov 25:ji2000887. doi: 10.4049/jimmunol.2000887. Epub ahead of print. PMID: 33239422.

[32] Ma Y, Liu S, Jun H, Wang J, Fan X, Li G, Yin L, Rui L, Weinman SA, Gong J, Wu J. A critical role for hepatic protein arginine methyltransferase 1 isoform 2 in glycemic control. FASEB J. 2020 Nov;34(11):14863-14877. doi: 10.1096/fj.202001061R. Epub 2020 Sep 12. PMID: 32918517.

[33] He L, Hu Z, Sun Y, Zhang M, Zhu H, Jiang L, Zhang Q, Mu D, Zhang J, Gu L, Yang Y, Pan FY, Jia S, Guo Z. PRMT1 is critical to FEN1 expression and drug resistance in lung cancer cells. DNA Repair (Amst). 2020 Nov;95:102953.

doi: 10.1016/j.dnarep.2020.102953. Epub 2020 Aug 20. PMID: 32861926.

[34] Li Z, Wang D, Lu J, Huang B, Wang Y, Dong M, Fan D, Li H, Gao Y, Hou P, Li M, Liu H, Pan ZQ, Zheng J, Bai J. Methylation of EZH2 by PRMT1 regulates its stability and promotes breast cancer metastasis. Cell Death Differ. 2020 Dec;27(12):3226-3242. doi: 10.1038/s41418-020-00615-9. Epub 2020 Sep 7. PMID: 32895488.

[35] Zheng D, Chen D, Lin F, Wang X, Lu L, Luo S, Chen J, Xu X. LncRNA NNT-AS1 promote glioma cell proliferation and metastases through miR-494-3p/PRMT1 axis. Cell Cycle. 2020 Jul;19(13):1621-1631. doi: 10.1080/15384101.2020.1762037. Epub 2020 May 18. PMID: 32420808; PMCID: PMC7469481.

[36] Jarrold J, Davies CC. PRMTs and Arginine Methylation: Cancer's Best-Kept Secret? Trends Mol Med. 2019 Nov;25(11):993-1009. doi: 10.1016/j.molmed.2019.05.007. Epub 2019 Jun 20. PMID: 31230909.

[37] Plotnikov A, Kozer N, Cohen G, Carvalho S, Duberstein S, Almog O, Solmesky LJ, Shurrush KA, Babaev I, Benjamin S, Gilad S, Kupervaser M, Levin Y, Gershovits M, Ben-Avraham D, Barr HM. PRMT1 inhibition induces differentiation of colon cancer cells. Sci Rep. 2020 Nov 18;10(1):20030. doi: 10.1038/s41598-020-77028-8. PMID: 33208761; PMCID: PMC7676271.

[38] Zhang P, Tao H, Yu L, Zhou L, Zhu C. Developing protein arginine methyltransferase 1 (PRMT1) inhibitor TC-E-5003 as an antitumor drug using INEI drug delivery systems. Drug Deliv. 2020 Dec;27(1):491-501. doi: 10.1080/10717544.2020.1745327. PMID: 32212935; PMCID: PMC7170320.